*Valim Levitin and Stephan Loskutov*
**Strained Metallic Surfaces**

## Related Titles

Herlach, D. M. (ed.)

**Phase Transformations in Multicomponent Melts**

2009
ISBN: 978-3-527-31994-7

Zehetbauer, M. J., Zhu, Y. T. (eds.)

**Bulk Nanostructured Materials**

2008
ISBN: 978-3-527-31524-6

Krupp, U.

**Fatigue Crack Propagation in Metals and Alloys**

Microstructural Aspects and Modelling Concepts
2007
ISBN: 978-3-527-31537-6

Levitin, V.

**High Temperature Strain of Metals and Alloys**

Physical Fundamentals
2006
ISBN: 978-3-527-31338-9

Hazotte, A. (ed.)

**Solid State Transformation and Heat Treatment**

2005
ISBN: 978-3-527-31007-4

Raabe, D., Roters, F., Barlat, F., Chen, L.-Q. (eds.)

**Continuum Scale Simulation of Engineering Materials**

Fundamentals – Microstructures – Process Applications
2004
ISBN: 978-3-527-30760-9

Westbrook, J. H., Fleischer, R. L. (eds.)

**Intermetallic Compounds**

Principles and Practice,
Volume 3: Progress
2002
ISBN: 978-0-471-49315-0

*Valim Levitin and Stephan Loskutov*

# Strained Metallic Surfaces

Theory, Nanostructuring and Fatigue Strength

WILEY-VCH Verlag GmbH & Co. KGaA

**The Authors**

*Prof. Dr. Valim Levitin*
Friedrich-Ebert-Str. 66
35039 Marburg
Germany

*Prof. Dr. Stephan Loskutov*
Lenin prospect 144 flat 41
69095 Zaporozyhe
Ukraine

All books published by **Wiley-VCH** are carefully produced. Nevertheless, authors, editors, and publisher do not warrant the information contained in these books, including this book, to be free of errors. Readers are advised to keep in mind that statements, data, illustrations, procedural details or other items may inadvertently be inaccurate.

**Library of Congress Card No.:**
applied for

**British Library Cataloguing-in-Publication Data**
A catalogue record for this book is available from the British Library.

**Bibliographic information published by the Deutsche Nationalbibliothek**
The Deutsche Nationalbibliothek lists this publication in the Deutsche Nationalbibliografie; detailed bibliographic data are available on the Internet at http://dnb.d-nb.de.

© 2009 WILEY-VCH Verlag GmbH & Co. KGaA, Weinheim

All rights reserved (including those of translation into other languages). No part of this book may be reproduced in any form – by photoprinting, microfilm, or any other means – nor transmitted or translated into a machine language without written permission from the publishers. Registered names, trademarks, etc. used in this book, even when not specifically marked as such, are not to be considered unprotected by law.

**Typesetting**  Uwe Krieg, Berlin
**Printing**  Strauss GmbH, Mörlenbach
**Binding**  Litges & Dopf GmbH, Heppenheim
**Cover Design**  Adam-Design, Weinheim

Printed in the Federal Republic of Germany
Printed on acid-free paper

**ISBN:** 978-3-527-32344-9

# Contents

**Introduction**  *1*

**1 Peculiarities of the Metallic Surface**  *7*
1.1 Surface Energy and Surface Stress  *7*
1.2 Crystal Structure of a Surface  *11*
1.3 Surface Defects  *14*
1.4 Distribution of Electrons near the Surface  *18*
1.4.1 Model of Free Electrons in Solids  *20*
1.4.2 Semi-Infinite Chain  *24*
1.4.3 Infinite Surface Barrier  *26*
1.4.4 The Jellium Model  *27*
1.5 Summary  *30*

**2 Some Experimental Techniques**  *33*
2.1 Diffraction Methods  *33*
2.1.1 The Low-Energy Electron Diffraction Method  *33*
2.1.2 The Reflection High-Energy Electron Diffraction Method  *40*
2.1.3 The X-ray Measurement of Residual Stresses  *40*
2.1.3.1 Foundation of the Method  *41*
2.1.3.2 Experimental Installation and Precise Technique  *44*
2.1.4 Calculation of Microscopic Stresses  *47*
2.2 Distribution of Residual Stresses in Depth  *47*
2.3 The Electronic Work Function  *48*
2.3.1 Experimental Installation  *50*
2.3.2 Measurement Procedure  *52*
2.4 Indentation of Surface. Contact Electrical Resistance  *53*
2.5 Materials under Investigation  *55*
2.6 Summary  *56*

*Strained Metallic Surfaces.* Valim Levitin and Stephan Loskutov
Copyright © 2009 WILEY-VCH Verlag GmbH & Co. KGaA, Weinheim
ISBN: 978-3-527-32344-9

## 3 Experimental Data on the Work Function of Strained Surfaces  59
3.1 Effect of Elastic Strain  *59*
3.2 Effect of Plastic Strain  *61*
3.2.1 Physical Mechanism  *65*
3.3 Influence of Adsorption and Desorption  *67*
3.4 Summary  *71*

## 4 Modeling the Electronic Work Function  73
4.1 Model of the Elastic Strained Single Crystal  *73*
4.2 Taking into Account the Relaxation and Discontinuity of the Ionic Charge  *76*
4.3 Model for Neutral Orbital Electronegativity  *78*
4.3.1 Concept of the Model  *78*
4.3.2 Effect of Nanodefects Formed on the Surface  *81*
4.4 Summary  *83*

## 5 Contact Interaction of Metallic Surfaces  87
5.1 Mechanical Indentation of the Surface Layers  *87*
5.2 Influence of Indentation and Surface Roughness on the Work Function  *93*
5.3 Effect of Friction and Wear on Energetic Relief  *95*
5.4 Summary  *100*

## 6 Prediction of Fatigue Location  103
6.1 Forecast Possibilities of the Work Function. Experimental Results  *104*
6.1.1 Aluminum and Titanium-Based Alloys  *104*
6.1.2 Superalloys  *107*
6.2 Dislocation Density in Fatigue-Tested Metals  *109*
6.3 Summary  *112*

## 7 Computer Simulation of Parameter Evolutions during Fatigue  115
7.1 Parameters of the Physical Model  *115*
7.2 Equations  *115*
7.2.1 Threshold Stress and Dislocation Density  *116*
7.2.2 Dislocation Velocity  *116*
7.2.3 Density of Surface Steps  *117*
7.2.4 Change in the Electronic Work Function  *117*
7.3 System of Differential Equations  *118*
7.4 Results of the Simulation: Changes in the Parameters  *118*
7.5 Summary  *120*

## 8 Stressed Surfaces in the Gas-Turbine Engine Components  123

8.1 Residual Stresses in the Surface of Blades and Disks and Fatigue Strength  123
8.1.1 Turbine and Compressor Blades  124
8.1.2 Grooves of Disks  126
8.2 Compressor Blades of Titanium-Based Alloys  128
8.2.1 Residual Stresses and Subgrain Size  130
8.2.2 Effect of Surface Treatment on Fatigue Life  133
8.2.3 Distribution of Chemical Elements  137
8.3 Summary  140

## 9 Nanostructuring and Strengthening of Metallic Surfaces. Fatigue Behavior  143

9.1 Surface Profile and Distribution of Residual Stresses with Depth  144
9.2 Fatigue Strength of the Strained Metallic Surface  150
9.3 Relaxation of the Residual Stresses under Cyclic Loading  154
9.4 Microstructure and Microstructural Stability  161
9.5 Empirical and Semi-Empirical Models of Fatigue Behavior  165
9.5.1 Fatigue-Crack Propagation in Linear Elastic Fracture Mechanics  166
9.5.2 Crack Propagation in a Model Crystal  171
9.6 Prediction of Fatigue Strength  173
9.7 Summary  178

## 10 The Physical Mechanism of Fatigue  181

10.1 Crack Initiation  181
10.2 Periods of Fatigue-Crack Propagation  192
10.3 Crack Growth  195
10.4 Evolution of Fatigue Failure  205
10.5 S – N curves  213
10.6 Influence of Gas Adsorption  215
10.7 Summary  216

## 11 Improvement in Fatigue Performance  219

11.1 Restoring Intermediate Heat Treatment  219
11.2 Effect of the Current Pulse on Fatigue  220
11.3 The Combined Treatment of Blades  223
11.4 Structural Elements of Strengthening  226
11.5 Summary  231

| | | |
|---|---|---|
| **12** | **Supplement I** *233* | |
| 12.1 | List of Symbols *233* | |
| 12.1.1 | Roman Symbols *233* | |
| 12.1.2 | Greek Symbols *235* | |
| | | |
| **13** | **Supplement II** *237* | |
| 13.1 | Growth of a Fatigue Crack. Description by a System of Differential Equations *237* | |
| 13.1.1 | Parameters to be Studied *237* | |
| 13.1.2 | Results *238* | |

**References** *243*

**Index** *247*

# Introduction

The properties of surface layers of solids are of fundamental importance in solid state theory.

A considerable body of evidence shows that atoms in surface layers exist under different conditions from the state of bulk atoms. It is the breaking of interatomic bonds near the surface that determines the special conditions on the surface. The arrangement of atoms at the surface is the key factor which affects the physical, mechanical and chemical properties of metallic materials and other solids, especially semiconductors and chemical compounds. The broken atom bonds cause distinct changes in the distribution of electrons.

On the other hand, surface layers play an important role in the behavior of solids in practical applications. Numerous technological processes influence the special properties of surface layers. Any strengthening or treatment of solids begins with the surface. An optimization of the surface microstructure allows one to produce materials with a nanocrystalline surface layer.

In turn, cracks, wear and rupture arise from the surface under normal operating conditions. Adsorption and chemical reactions also begin on the surface.

Fatigue rupture and the high-temperature creep originate in surface layers. The structure of the surface is responsible for fatigue strength, corrosion resistance, and wearing capacity. Surface layers contain defects of different dimensions, e.g. vacancies, dislocations, distortions, steps, adsorbed atoms.

Over recent decades, physics has made significant progress in the study of surfaces of metals, alloys, and semiconductors on a microscopic scale.

It is helpful to make the study of surfaces as simple as possible by eliminating extraneous factors. From the physical point of view it is necessary to investigate an ideal clean surface.

In order to keep the surface fairly clean specimens need to be in an ultrahigh vacuum. One should keep the residual gas pressure lower than $10^{-8}$ Pa ($7.5 \times 10^{-11}$ Torr) for a long time. Vacuum physics and technology gave a strong impetus to surface science. The preparation of well-defined, clean surfaces, which are usually investigated, became possible only after the development of the ultra-high vacuum technique.

*Strained Metallic Surfaces.* Valim Levitin and Stephan Loskutov
Copyright © 2009 WILEY-VCH Verlag GmbH & Co. KGaA, Weinheim
ISBN: 978-3-527-32344-9

However, the majority of industrial processes occur in the atmosphere. Real parts and components operate under usual atmospheric conditions. Most scientists and engineers do not find an ideal and absolutely pure solids surface. Deformation distortions, chemical impurities, and absorption layers are typical for solid surfaces, in reality. These facts substantiate the investigation of surfaces in non-ideal conditions.

The main goal of our book is the study of strained metallic surfaces under ordinary conditions.

A strained state of the metallic surface is the outcome of prior treatment and manufacture.

Residual stresses are produced by plastic deformation, thermal contractions or can be induced by a production process. The residual mechanical stresses are known to balance in the macroscopic and microscopic areas of materials. The plastic strain of the surface is well-known in the industry as a way to improve the fatigue and strength properties of metallic parts. Shot peening, deep rolling, hammering, treatment by metallic balls in an ultrasonic field, and laser shock peening are the new technological methods used to increase the strength of crucial parts. All these processes somehow or other induce the near-surface nanostructures. The generation of nanostructured surface layers is expected to improve the properties of materials. Structural changes in the near-surface regions of metals and alloys are of great interest and have not yet been sufficiently investigated.

The electronic work function is of special and increasing interest to material scientists and engineers because of its sensitivity to the physical state of the surface. Metals are known to consist of two subsystems. These are the relatively slow crystal lattice of ions and the gas of the fast free electrons. External and residual stresses result in considerable changes in the ionic crystal lattice.

The properties of electron emission indicate close interaction between ionic and electronic subsystems. It is essential, however, to emphasize that a relationship between the emission properties and elastic and plastic strain is not yet sufficiently understood. Here we would like to make up for this deficiency of data.

The vast development of surface science has been covered in many excellent books on the technique of surface investigation, surface structure, surface processes, and the theoretical modeling of the surface. Previous investigations and books have deepened our knowledge on the problem and stimulated much experimental work. However, there is a gap in the existing literature. It is preferable to study pure metals in theoretical investigations. The physical fundamentals of surface deformation, especially of industrial nickel-based superalloys and titanium-based alloys, are not clearly understood.

We believe there is a need for a book to act as a bridge between a theory and its practical applications. This book is an attempt to bridge the gap between surface physics and physics of solids and technology. Our goal is to consider physical theories as well as the applied aspects of the strained surface problem.

The book treats data from systematic experimental measurements of important characteristics which are related to physical fundamentals of peculiarities of strained metallic surfaces. The book is designed to cover data accumulated during recent decades when studying the properties of these surfaces.

The lower layers of the solid surface are involved in deformation processes, as well as the top ones. In this book the metallic surface is regarded as the top of tens of atomic layers.

The book consists of eleven chapters.

A succinct description of the features of the metallic surface is presented in the first chapter. We recall the concepts of surface energy and surface stress. The defects at the surface are discussed. We describe the crystal structure of the surface and the distribution of free electrons near the surface. We would also like to remind the reader of the quantum phenomena related to the surface.

The second chapter is devoted to some techniques for experimental studies of the strained metallic surfaces. Diffraction methods are considered. We describe the technique of precise X-ray measurements of residual macroscopic stresses. A new installation and the method of measurement of the electronic work function are presented. Attention is given also to mechanical methods of studying of the surface layers. The materials under investigation are described.

The emission properties of the strained metallic surface are the subject of the third chapter. A scanning Kelvin probe method is used to investigate the strained and stressed surface of metals and alloys. Data on the response of the electronic work function to the elastic and plastic deformation of metals are presented and discussed. The physical mechanism of processes is deduced from the results obtained and the phenomena responsible for the variation in the work function are discussed.

The fourth chapter deals with an examination of various models of the electronic work function for the strained surface. Some theoretical models are proposed. Equations are derived for the calculation of the electronic work function of the imperfect surface and methods of calculation of the work function of the elastic strained metal surface, are developed. A theoretical basis for this consideration is a self-consistent scheme of the work function calculation which took into account essential corrections to the jellium model as well as the formation of nanometric surface defects.

In the fifth chapter we consider the contact interaction of metallic surfaces which is inherent in processes of wear and tear, mechanical treatment and strengthening. The technique of local indentation which is one of the effective methods of measurement of properties of the materials considered. The work function procedure is found to be sensitive to processes of friction and wear.

Data on the fatigue prediction for metals and alloys are presented in the sixth chapter. The basic mechanism of fatigue fracture is the origination of a crack on the surface and a slow propagating of the crack. Fatigue of materials is known to be a dangerous phenomenon. We discover that the nondestructive method of work function measurement may be used to predict the initiation of fatigue cracks.

The seventh chapter contains a computer stimulation of the evolution of structure parameters during fatigue. We work out a physical model that describes processes leading to fatigue. Our approach is to derive a system of ordinary differential equations and to solve the system numerically.

The eighth chapter deals with data on surface residual stresses and fatigue life of gas-turbine components. The need to use high stresses in aircraft causes initiation of fatigue cracks. The induced favorable residual stresses are used in industry to increase the fatigue strength. The authors present their results of the study of gas-turbine blades and discs. In this chapter we consider the distribution of induced macroscopic residual stresses for real engine parts and the effect of surface treatments on the fatigue strength. Microscopic stresses and subgrain sizes are also measured.

The ninth chapter contains a considerable body of evidence on nanostructuring and strengthening of metallic surfaces. Various mechanical treatments of the surface and induced nanometric structures are considered. Favorable structural factors which have a significant influence on fatigue life and lead to additional increase in fatigue strength are also discussed. We describe the distribution of residual stresses with the depth, their evolution during fatigue tests, and the stability of the microstructure during cycling. Semi-empirical models of fatigue behavior are presented.

In the tenth chapter we derive a quantitative physical mechanism for fatigue. The fatigue phenomenon is considered on the atomic scale. We discover that fatigue damage is, at first, reversible. The initiation and propagation of fatigue cracks are under study. We place an emphasis on the role of the crack growth of vacancy flow and of the stress gradient near the crack tip. Periods of the fatigue damage are considered from the physical point of view. A focus is placed on the dependence of the crack growth rate on the cycling time and the stress amplitude. We derive equations for the assessment of embryo crack length, and number of cycles until specimen fracture.

In the eleventh chapter we consider some new methods of fatigue life prolongation. Topics that are discussed concern the intermediate thermal treatment, the processing of alloys by electric impulses, and a combined strengthening of compressor blades. This consists of the restoring of components, the severe plastic strain of the surface, and the armoring of blades by a coating. We discuss also structure elements of fatigue performance.

A detailed review of all aspects of the problem under consideration for a pure ideal surface goes beyond the scope of this book. Therefore, the known principles and established facts are mentioned only briefly. The reader can find reviews concerning the physics of ideal surfaces in different books and articles, for example [1-8].

This book is intended for students and postgraduate students in the area of solid state physics, surface physics, materials engineering, and physical chemistry. At the graduate and postgraduate level there is reason to believe that the book will meet the needs of those concerned with the properties, investigation and application of modern industrial alloys.

We hope that the book will also be useful for material scientists, engineers, researchers and practitioners from the industry sectors who are interested in problems of material properties, surface strengthening, nanostructuring, fatigue strength, and physicochemical activity.

# 1
# Peculiarities of the Metallic Surface

## 1.1
### Surface Energy and Surface Stress

The first law of thermodynamics states that

$$dQ = dU + dW \tag{1.1}$$

where $dQ$ is the increment in the heat energy of the system, $dU$ is the increment in the internal energy and $dW$ is a work that the system has performed. Only $dU$ is the total differential and it is independent of the method of system transition from one state to another.[1] $dQ$ and $dW$ are simply infinitesimal quantities. In fact, (1.1) is the law of the energy conservation for an isolated system.

Substituting $dQ = T\,dS$ and $dW = p\,dV$ in (1.1) one arrives at

$$dU = T\,dS - p\,dV \tag{1.2}$$

where $S$ is the entropy, $T$ is the temperature, $p$ is the pressure, $V$ is volume of the system.

The internal energy of the system depends also on the number $N$ of particles, that is, atoms or molecules. Thus, one should add the corresponding term:

$$dU = T\,dS - p\,dV + \mu\,dN \tag{1.3}$$

where $\mu$ is the chemical potential. The chemical potential is defined as a change in the internal energy when the number of particles varies at constant entropy and volume:

$$\mu = \left(\frac{\partial U}{\partial N}\right)_{S,V} \tag{1.4}$$

---

[1] If the internal energy is a function of the pressure and the temperature, $U = U(p, T)$, then the infinitesimal increment in the internal energy is given by $dU = \left(\frac{\partial U}{\partial p}\right)_T dp + \left(\frac{\partial U}{\partial T}\right)_p dT$.

*Strained Metallic Surfaces.* Valim Levitin and Stephan Loskutov
Copyright © 2009 WILEY-VCH Verlag GmbH & Co. KGaA, Weinheim
ISBN: 978-3-527-32344-9

The internal energy $U$ increases (or, on the contrary, decreases) if
- the system receives (returns) the heat or
- mechanical work is done under the system (the system does mechanical work) or
- the number of particles in the system increases (decreases).

Turning to the free energy. The Helmholtz free energy $F$ is known to be equal to

$$F = U - TS \tag{1.5}$$

Differentiating (1.5) and combining with (1.3) we obtain

$$dF = -S\,dT - p\,dV + \mu\,dN \tag{1.6}$$

At constant temperature and volume the increment in the free energy varies linearly with increase in the particle number,

$$dF = \mu\,dN \tag{1.7}$$

with the proportionality factor $\mu = \mu(T, V)$.

The Gibbs thermodynamic potential $G = F + pV$ is expressed as

$$dG = -S\,dT + V\,dp + \mu\,dN \tag{1.8}$$

The difference between the Helmholtz free energy and the Gibbs thermodynamic potential is insignificant under atmospheric pressure for a bulk solid.

Molecules of the surface layer in a homogeneous liquid are known to be attracted by other molecules within the liquid. Unlike in the body of the liquid the attraction of surface molecules is not compensated. Liquids turn out to be covered by an elastic stretched film and so the concept of surface tension is used for them.

The thermodynamic approach can also be applied to the surface of solids. J. W. Gibbs was the first to note that the surface contributes to the free energy. He also considered the cleavage process of a bulk body.

In order to create a free surface one must break the bonds between neighboring atoms. This implies that the creation of an additional piece of surface costs the system extra energy. The surface energy is equal to the work necessary in order to form the unit area of surface by a process of division of the solid into parts. This process is assumed to be thermodynamically reversible.

The reversible work $dW$ required for an external force to create an infinitesimal area $dA$ of the surface is directly proportional to this area

$$-dW = \gamma\,dA \tag{1.9}$$

where $\gamma$ is a proportionality factor. It is dependent on temperature, volume, and the particle number.

The minus in (1.9) implies that the work performed by an external force is expended for increase in the Helmholtz free energy of $dF$. Since, for a reversible and equilibrium process, $dF = -dW$, the surface contribution to the free energy must be proportional to the increment in the surface area,

$$dF = \gamma \, dA \tag{1.10}$$

The factor of proportionality $\gamma$ can be identified as the excess of the free energy per unit area. Thus, the surface energy $\gamma$ is also defined as

$$\gamma = \left(\frac{dF}{dA}\right)_{T,V,N} \tag{1.11}$$

The temperature, the volume of crystal and the number of atoms are assumed to be constant. The reversibility requirement in the definition of $\gamma$ implies that the composition in the surface region is also in thermodynamic equilibrium. Recall also that distances between atoms remain unchanged.

The unit of surface energy is $J\,m^{-2}$ or $N\,m^{-1}$.

A variation in the free energy of a solid through the equilibrium cleaving process, in a general case, is given by

$$dF = -S\,dT - p\,dV + \mu\,dN + \gamma\,dA \tag{1.12}$$

Gibbs pointed out that, for solids, there is another important quantity to be considered. It is the work per unit area needed to elastically stretch (or compress) a pre-existing surface. Solid surfaces exhibit a surface stress which is similar to surface tension in a liquid.

As long ago as 1950 Shuttleworth [9] emphasized the distinction between the surface Helmholtz free energy, and the surface stress. The surface energy is the work necessary to form unit area of a surface by a process of splitting up, the cleavage. The surface stress is the tangential stress (force per unit length) in the surface layer; this stress must be balanced either by external forces or by volume stresses in the body. For crystals, the surface stress is not equal to the surface energy. Both quantities have the same units, but the surface stress is a second-rank tensor, whereas the surface energy is a scalar quantity.

The broken interatomic bonds cause the surface stress. The atomic configuration at the surface is dissimilar to that within the body. Surface atoms therefore have a different arrangement than if they were constrained to remain in the interior of the solid. Therefore, the interior atoms are viewed as exerting a stress on the surface (moving them out of the positions they would otherwise occupy).

The cleavage of a body creates new surfaces. The excess in the surface free energy $dF_s$ consists of two parts,

$$dF_s = dF_{s1} + dF_{s2} \tag{1.13}$$

or

$$dF_s = d(\gamma A) = \gamma\, dA + A\, d\gamma. \tag{1.14}$$

The first term $dF_{s1} = \gamma\, dA$ describes the work needed to break down interatomic bonds and to create the surface area $dA$. The average area per atom is unchanged, but the number of atoms on the surface increases.

The second term $dF_{s2} = A\, d\gamma$ contributes the free energy due to changes in the interatomic distances on new surfaces. These changes are balanced in the macroscopic and microscopic areas. The number of atoms on the surfaces remain unchanged. The stressed surface contributes a surface energy with an increment $d\gamma$.

A change in the surface area may be considered in a macroscopic sense as the consequence of stresses acting on surface atoms [6]. A tensor with components $\sigma_{ij}$ (where $i, j = 1, 2$) defines the stresses on the surface in an approximation of elasticity theory. The strain tensor has components $\varepsilon_{ij}$. It is used to express the elastic deformation of the solid surface.

Consider a plane normal to the surface and label the normal to the plane as the direction $j$. $\sigma_{ij}$ is the force per unit length which atoms exert across the line of intersection of the plane with the surface in the $i - th$ direction. The strain tensor $\varepsilon_{ij}$ is defined in a similar way.

The work done in the surface can be expressed as the change in the elastic energy

$$dF_s = A \sum_{i,j=1}^{2} \sigma_{ij} d\varepsilon_{ij} \tag{1.15}$$

The increment in the surface is given by

$$dA = A \sum_{i=1}^{2} d\varepsilon_{ii} = A \sum_{i,j=1}^{2} \delta_{ij} d\varepsilon_{ij} \tag{1.16}$$

The total differential of the surface stress $\gamma$ is written as

$$d\gamma = \sum_{i,j=1}^{2} \frac{\partial \gamma}{\partial \varepsilon_{ij}} d\varepsilon_{ij} \tag{1.17}$$

Combining (1.14)–(1.17) and making the coefficients at $d\varepsilon_{ij}$ equal one obtains

$$\sigma_{ij} = \gamma \delta_{ij} + \frac{\partial \gamma}{\partial \varepsilon_{ij}} \tag{1.18}$$

The surface stress $\sigma_{ij}$ and the stress-induced elastic strain are tensors related to each other by (1.18).

For most surfaces, the off-diagonal components are equal to zero in appropriate coordinates, that is, the surface stress is isotropic. In this case, the surface stress can be viewed as a scaler quantity

$$\sigma = \gamma + \frac{d\gamma}{d\varepsilon} \qquad (1.19)$$

This indicates that the difference between the surface free energy and the surface stress is the change in the free energy per unit change in the elastic strain of the surface. A better way is to consider $\sigma$ as the force per unit length (or work per unit area) exerted on a surface during elastic deformation; and to consider $\gamma$ as the force per unit length (or work per unit area) exerted on a surface during plastic deformation [6].

The surface energy depends on the crystal plane orientation. The value $\gamma$ is a function of $\vec{n}$, $\gamma = f(\vec{n})$, where $\vec{n}$ is perpendicular to a plane $(hkl)$ in the crystal lattice.

The experimental measurement of the surface energy for solids is extremely difficult. There are few data about values of $\gamma$ in the literature. However, several efforts to calculate the surface energy have been made.

The authors of [11] used the density functional theory to establish a database of surface energies for low index surfaces of 60 metals from the periodic table. The absolute values are of the order of several $J m^{-2}$. The values vary when the surface orientation changes.

Some data are presented in Table 1.1.

**Table 1.1** Surface energies for different crystal planes in some metals, $J m^{-2}$.

| Metal | (100) | (110) | (111) | Ref. |
| --- | --- | --- | --- | --- |
| Al | 1.35 | 1.27 | 1.20 | 11 |
| Au | 1.63 | 1.70 | 1.28 | 11 |
| Mo | 3.34 | 2.92 | 3.24 | 10 |
| W | 4.64 | 4.01 | 4.45 | 11 |

## 1.2
## Crystal Structure of a Surface

A scheme of some characteristic rearrangements of surface atoms for the cubic crystal lattice is presented in Figure 1.1. A compression or an extension of the topmost layer leads to change in interatomic spacing. The bulk spacing $c_{bulk}$ decreases to values $c_2$ and $c_1$. This variation is normal to the surface and is called the relaxation (Figure 1.1a).

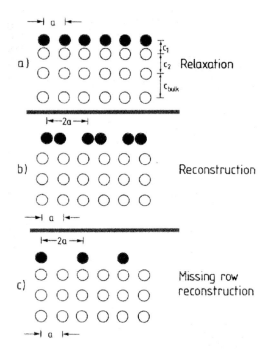

**Fig. 1.1** Schematic view of characteristic rearrangements of surface atoms in a cubic crystal lattice. $a$ is the crystal lattice constant. Reprinted from [4] with permission from Springer.

The atomic configuration shown in Figure 1.1b is related to a shift in the atoms parallel to the surface. This type of atomic rearrangement is called reconstruction. Missing row reconstruction is also possible (Figure 1.1c).

In Figure 1.1 only idealized models of atom distributions are shown. In reality, the atomic rearrangements are more complex. The atom structure of the surface is dependent on the bond type. It is different in semiconductors with their directed covalent bonds and in metals, which have delocalized electronic bonds. In many semiconductors, which are tetrahedrally bonded, such as silicon, germanium, GaAs, and InP, the directedness of bonds in the bulk has a dramatic effect and may result in directed bonds at the surface.

There are no convincing experimental data on surface relaxation [1]. Calculations on the relaxation of the surface layer for metallic crystals have been made using models in which the bonds between pairs of atoms were considered.

The results for copper obtained using a Morse interaction potential are shown in Figure 1.2. It is found that the surface layer is displaced by 5–20%. The displacement of atoms from equilibrium positions decreases exponentially with the layer number. The greater the atom density in the crystal plane the less atomic displacements there will be in the surface layers.

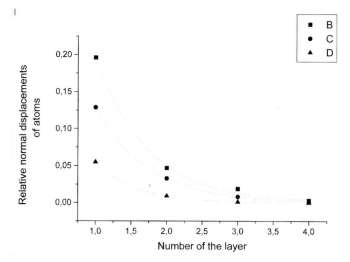

**Fig. 1.2** Displacements of atom layers near the surface of copper. B, crystal plane (110), the atom density equals 10.86 at. $nm^{-2}$; C, plane (100), 15.36 at. $nm^{-2}$; D, plane (111), 17.74 at. $nm^{-2}$. The calculated data from [12] are plotted in the graph.

The structure of the first crystal plane differs from that of the underlying planes even in an ideal clean surface. Such a structure is in fact a superstructure (the overlayer structure). One uses a special notation in order to designate the dimensions, nature and orientations of the unit cell of the first plane relative to the underlying ones. These notations are useful also when an ordered adsorbed layer is present at the surface.

The position of an atom at the surface is given by the vector $\vec{r}$

$$\vec{r} = m\vec{a} + n\vec{b} \tag{1.20}$$

where $\vec{a}$ and $\vec{b}$ are the translation vectors of the solid and $m$ and $n$ are integers. Then the superstructure can be labeled by E $\{hkl\}p(m \times n)$ or E $\{hkl\}c(m \times n)$. E is a chemical element, $p$ represents a primitive cell, $c$ denotes a centered one. The surface unit cell can be rotated with respect to the substrate unit cell. The ratio of the dimensions of these cells cannot be an integer. The notation $E\{hkl\}c(m \times n)R\theta$ means a surface structure that is obtained from the surface plane unit cell by a rotation trough an angle $\theta$, the length of the basis vectors being multiplied by $m$ and $n$, respectively.

For instance, a clean surface {100} for nickel is denoted by $Ni\{100\}c(1 \times 1)$. The superstructure formed by the absorbtion of oxygen atoms on this surface is the $O\{100\}c(2 \times 2)$ structure.

Matrix notation is also used for the occurring superstructure.

The relationship between the vectors of an overlayer structure $\vec{a_2}, \vec{b_2}$ and basis vectors $\vec{a_1}, \vec{b_1}$ that correspond bulk layers is given by

$$\vec{a_2} = G_{11}\vec{a_1} + G_{12}\vec{b_1}; \tag{1.21}$$

$$\vec{b_2} = G_{21}\vec{a_1} + G_{22}\vec{b_1}. \tag{1.22}$$

Thus, vectors $\vec{a_2}, \vec{b_2}$ of an overlayer structure can be obtained from basis vectors $\vec{a_1}, \vec{b_1}$ by the conversion

$$\begin{pmatrix} G_{11} & G_{12} \\ G_{21} & G_{22} \end{pmatrix} \tag{1.23}$$

Figure 1.3 presents an example of the notations.

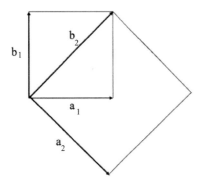

**Fig. 1.3** Vectors $\vec{a_2}$ and $\vec{b_2}$ define the superlattice, vectors $\vec{a_1}$ and $\vec{b_1}$ are the vectors of the plane inside the bulk. The surface superstructure has a possible notation E $\{100\}c(\sqrt{2} \times \sqrt{2})\ R45°$ or $\begin{pmatrix} 1 & \bar{1} \\ 1 & 1 \end{pmatrix}$.

In such a case

$$\vec{a_2} = \vec{a_1} - \vec{b_1};$$

$$\vec{b_2} = \vec{a_1} + \vec{b_1}.$$

Consequently, the corresponding matrix is

$$\begin{pmatrix} 1 & \bar{1} \\ 1 & 1 \end{pmatrix} \tag{1.24}$$

## 1.3
### Surface Defects

Defects always exist on real surfaces. Figure 1.4 shows, schematically, defects of different dimensionality. Defects of zero dimension are vacancies, adatoms, ledge adatoms, and kinks. Their sizes in three dimensions are of the order of the interatomic distance. On the surface of a compound crystal one can distinguish between adatoms of the same kind or foreign adatoms. The step is a one-dimensional defect, in which the ledge separates two terraces from each other. In many cases steps of the single atomic height prevail over steps of several atomic heights. There are also groups of defects such as divacancies, and steps of some interatomic distances in height.

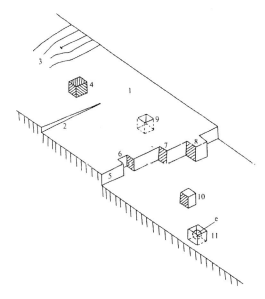

**Fig. 1.4** Various defects that may occur on a solid surface: 1, a terrace, i.e. the perfect flat face itself; 2, an emerging screw dislocation; 3, the intersection of an edge dislocation with the terrace; 4, an impurity adatom; 5, a monoatomic step; 6, a vacancy in the ledge; 7, a step in the ledge–a kink; 8, an adatom of the same kind as the bulk atoms situated upon the ledge; 9, a vacancy in the terrace; 10, an adatom on the terrace; 11, a vacancy in the terrace where an electron is trapped. Reprinted from [2] with permission from Oxford University Press.

Different positions of atoms have various numbers of neighboring atoms. For example, an atom on a terrace has the largest number of neighbors. Atoms on steps and adatoms have the least number of neighbors.

Other important defects are dislocations which can be generated by the strained surface. (Figure 1.5). An edge dislocation penetrating into a surface with the Burgers vector oriented parallel to the surface, creates a point defect. In a typical well-annealed single crystal the density of dislocation–surface interactions is of the order of $10^8 - 10^{10}$ m$^{-2}$.

Formation of steps at the surface is shown schematically in Figure 1.6. The sources of the dislocations are situated along slip lines. Dislocations emitted by sources move along the slip planes and appear on the surface.

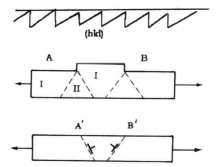

**Fig. 1.5** Generation of dislocations at a surface. The surface roughness on a plane $(hkl)$ is shown. During deformation the surface acts as a dislocation source. $A'$ and $B'$ are slip planes, $A$ and $B$ are steps that have opposite signs, $I$ and $II$ are shifted parts of the crystal.

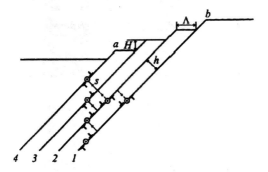

**Fig. 1.6** Formation of steps at the metallic surface. $ab$ is a slip band, 1 - 4 are slip lines, $s$ are dislocation sources, $h$ is the mean distance between slp lines, $H$ is the mean hight of steps.

The surface is known to attract a dislocation with the force $F = -\mu b^2/4\pi l$, where $\mu$ is the shear modulus, $b$ is the Burgers vector and $l$ is the distance from the free surface. A step is formed at the surface. Many dislocations take part in the formation of a step. However, polyatomic steps are unstable energetically and transform to monatomic steps.

The microscopic structure of the surface is changed as a result of deformation. The technique of tunnel microscopy enables one to reveal the formation and change in the nanometric defects on the surface of thin polished films of copper [13].

The applied stress leads to the generation of four ensembles of surface nanometric defects. These ensembles differ from one another in dimension and energy of formation. They have the shape of a prism. One of the walls is perpendicular to the surface, another one is inclined at 30°. Figure 1.7 shows an example of a nanometric surface defect. Walls of nanodefects consist of steps with a width of from 5 to 50 nm. Their walls are formed as a consequence of the output of dislocations along planes of the light slip. Sources of dislocations belong also to a set of four ensembles. The authors [13] believe that when the concentration of nanodefects amounts to $\sim 5\%$ part of them resolves. Another part enlarges into defects of the next ensemble.

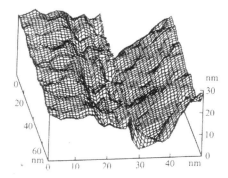

**Fig. 1.7** The nanometric defect on the surface of strained copper (after [13]).

Steps are important in the formation of vicinal surfaces, that is, high-index surfaces. Such vicinal surfaces are formed by small low-index terraces and a high density of regular steps. The steps in a simple cubic structure are shown in Figure 1.8. The normal to the surface is slightly inclined to the [001] direction. The inclination angle $\vartheta$ to the [001] axes is given by $\tan \vartheta = 1/4$.

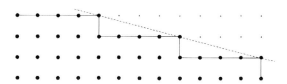

**Fig. 1.8** Terraces, ledge atoms and steps in a simple cubic crystal. Terraces are located along (001) plane, steps are parallel to (010). The surface plane is (014). Reprinted from [6] with permission from Springer Science.

A convenient notation was proposed [14] for the geometrical structure of steps. A stepped surface is denoted by

$$p(hkl) \times (h'k'l'),$$

where $hkl$ are the Miller indices of the terraces, $(h'k'l')$ are the indices of the ledges, and $p$ gives the number of atomic rows in the terrace parallel to the edge. In Figure 1.9a 6(111) × (001) stepped surface is presented. A series of six-atom terraces are separated by (100) × (111) steps in the face-centered cubic crystal lattice.

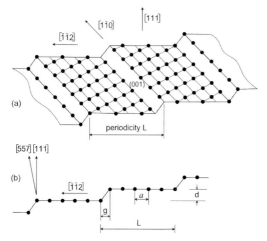

**Fig. 1.9** Steps on the (557) surface of a face-centered cubic crystal. (a) Positions of the lattice points in a 6(111) × (011) stepped surface. (b) The characteristic length and heights are shown. Reprinted from [6] with permission from Springer Science.

This type of stepped surface is obtained by cutting the crystal along a plane making a small angle ($\leq 10°$) with a low-index plane. Such surfaces show a periodic succession of steps of monoatomic height and flat terraces.

On metal surfaces the step tends to be smoothed out by the gas of free electrons, which forms dipole moments due to the spatially fixed positive ion cores.

In semiconductors the different dangling (free) bonds might modify the electronic energy levels near the steps.

Surface defects are centers of chemical activity. The degree of adsorption increases exponentially with an increase in the defect density.

The resistance of the material to deformation is strongly affected by films on the surface. The scheme in Figure 1.10 is an illustration of the decrease in strength of a crystal with an active liquid on the surface (the Rebinder effect). Other surface films can increase the strength.

## 1.4
**Distribution of Electrons near the Surface**

Many microscopic phenomena on the surface result from special features of the electron subsystem. Properties of the electronic structure near the surface

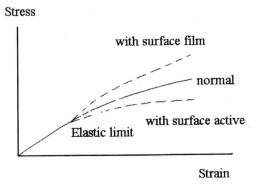

**Fig. 1.10** The dependence of stress on strain for the different surface states.

are somewhat different from those inside the bulk. This is mainly because part of interatomic bonds are broken at the surface. These broken bonds are called dangling bonds. This means that unpaired electrons tend to bond with each other or with foreign atoms. The neighboring atoms can form pairs that are called dimers.

In Figure 1.11 the bonds between the surface atoms are shown. The lattice parameter of the surface crystal cell increases to twice its length in one direction, compared with the perpendicular direction. This effect is termed the $2 \times 1$ reconstruction.

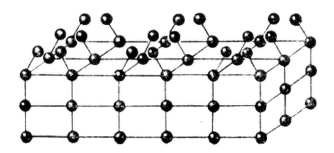

**Fig. 1.11** Formation of atomic dimers on the surface. Reconstruction $2 \times 1$.

There is still no reliable experimental technique for the unequivocal determination of the atomic structure of the topmost surface layer. One has therefore to consider various models of the surface structure. As usual, the calculations based on the model have to be compared with experimental data.

## 1 Peculiarities of the Metallic Surface

A quantum-mechanical description of the electronic properties has been worked out for simple metals. It begins with the description of a gas of free electrons. In order to describe the basic properties of the surface one examines the problem of many bodies in the framework of density functional theory. This theory was developed in order to discuss the basic electronic properties of metals.

Here we follow the approach of Blakely [1] as well as that of Lüth [4] and Kiejna and Wojciechowsky [5].

### 1.4.1
### Model of Free Electrons in Solids

Let us consider a cubic box of size $L \times L \times L$ of a monovalent metal. Inside the metal crystal the electrostatic potential $V(\vec{r})$ is a periodic function of the distance. However, electrons are trapped by the metallic ions inside the metal. According to the Sommerfeld idea one can represent the bonding between ions and electrons by an infinite potential barrier at the metal surface. In that case the electrostatic potential $V(\vec{r}) = $const$ = a$ in the box and $V(\vec{r}) = \infty$ elsewhere. The positive charge is distributed inside the cubic box with a constant density $\varrho_+(r) = ne$, where $n$ is the number of ions per unit volume and $e$ is the charge of the electron. The state of each electron is described by the Schrödinger wave equation

$$\left[-\frac{\hbar}{2m}\nabla^2\psi + V\psi\right] = E\psi \qquad (1.25)$$

where $\hbar = h/2\pi$, $h$ is the Planck constant, $m$ is the mass of an electron, $\psi$ is the wave function, $V$ is the potential due to all the other electrons and the positive charge in the system and $E$ is the total energy of the electron[2].

The motion of the electron is limited by the surface. The energy $E$ of the electron in a bound state varies discretely.

The wave function squared, $\psi^2$, determines the local position of the electron in space. For example, in the one-dimensional case the probability $w$ to find an electron in a segment $dz$ is given by

$$w(z) = \psi^2(z)\, dz$$

The function $w(z)$, as any probability function, has a property

$$\int_{-\infty}^{+\infty} \psi^2(z)\, dz = 1$$

---

[2] Atomic units are used in quantum physics. The Bohr radius $a_0 = 0.529177 \times 10^{-10}$ m is a unit of length. The mass of the electron is equal to $m = 9.109534 \times 10^{-31}$ kg. The electron charge $e = 1.602169 \times 10^{-19}$ C, and the Planck constant $\hbar = 1.054589 \times 10^{-34}$ J s. In order to convert SI units into atomic units one has to put $m = \hbar = e = 1$.

## 1.4 Distribution of Electrons near the Surface

The boundary conditions $\psi(x,y,z) = 0$ must be justified on the surfaces of the box under consideration. The solution of (1.25) is given by

$$\psi(x,y,z) = \left(\frac{8}{L^3}\right)^{1/2} \sin\left(\frac{\pi n_x x}{L}\right) \sin\left(\frac{\pi n_y y}{L}\right) \sin\left(\frac{\pi n_z z}{L}\right) \quad (1.26)$$

where positive integers $n_x, n_y, n_z = 1, 2, 3, \ldots$. On the surface of the cube, the boundary conditions $\psi(x,y,z) = 0$ are fulfilled.

The energies of the electron can have only discrete quantities. They are given by

$$E = \frac{\pi^2 \hbar^2}{2mL^2}\left(n_x^2 + n_y^2 + n_z^2\right) \quad (1.27)$$

The possible kinetic energies of the electron can also be expressed as

$$E = \frac{\hbar}{2m}\vec{k}^2 = \frac{\hbar}{2m}(k_x^2 + k_y^2 + k_z^2) \quad (1.28)$$

where $k_x, k_y, k_z$ are components of the wave vector $\vec{k}$:

$$k_x = \frac{\pi n_x}{L}$$

and similarly for $k_y$ and $k_z$.

The lowest energy level is equal to

$$E(1,1,1) = \frac{\pi^2 \hbar^2}{2mL^2}(1+1+1) \quad (1.29)$$

Only two electrons can have the same energy level. According to the Pauli principle, successive electrons have to occupy higher levels. The energy of the electrons of the highest occupied state at $T = 0$ is called the Fermi energy $E_F$. It is related to the concentration of electrons:

$$E_F = \frac{\hbar^2}{2m}(3\pi^2 \bar{n})^{2/3} \quad (1.30)$$

where $\bar{n}$ is the average electron density. The available energetic states are filled at $T = 0$ K up to a maximum value of $|\vec{k}|$. This is denoted by $k_F$ and is called the Fermi momentum.

$$k_F = (3\pi^2 \bar{n})^{1/3} \quad (1.31)$$

Thus

$$E_F = \frac{\hbar^2}{2m} k_F^2 \quad (1.32)$$

A characteristic volume falling on one electron is equal to

$$\frac{4}{3}\pi r_0^3 = \frac{L^3}{N} \tag{1.33}$$

In units of the Bohr radius $a_0$ the so called Wigner–Seitz radius $r_s$ is given by

$$r_s = \frac{r_0}{a_0} = \frac{(3/4\pi\bar{n})^{1/3}}{a_0} \tag{1.34}$$

where $r_s$ is measured in atomic units.

In Table 1.2 parameters of the electron subsystem in some metals are presented. The parameters were calculated based on the valency of metals and volumes of their crystal unit cells.

**Table 1.2** Parameters of the electron subsystem in some metals. $Z$ is the number of the element in the periodic table. fcc is the cubic face-centered crystal lattice, bcc is the cubic body-centered, hcp is the hexagonal close-packed, cc is the cubic complicated crystal lattice. $\bar{n}$ is the electron concentration, $E_F$ is the Fermi energy, $r_s$ is the characteristic radius, $k_F$ is the Fermi momentum.

| Metal | Z | Crystal lattice | Valency | $\bar{n}, 10^{29}$ m$^{-3}$ | $E_F$, eV | $r_s$, a.u. | $k_F, 10^{10}$ m$^{-1}$ |
|---|---|---|---|---|---|---|---|
| Na | 11 | bcc | 1 | 0.254 | 3.15 | 3.99 | 1.75 |
| Mg | 12 | hcp | 2 | 0.867 | 7.14 | 2.65 | 1.37 |
| Al | 13 | fcc | 3 | 1.81 | 11.7 | 2.07 | 1.75 |
| Ti | 22 | hcp | 4 | 3.02 | 16.5 | 1.75 | 2.08 |
| V | 23 | bcc | 5 | 3.58 | 18.4 | 1.65 | 2.20 |
| Cr | 24 | bcc | 6 | 5.03 | 23.1 | 1.47 | 2.46 |
| Mn | 25 | cc | 2 | 1.65 | 10.9 | 2.14 | 1.70 |
| Fe | 26 | bcc | 2 | 1.70 | 11.1 | 2.12 | 1.71 |
| Co | 27 | hcp | 2 | 4.82 | 22.7 | 1.50 | 2.43 |
| Ni | 28 | fcc | 2 | 1.84 | 11.8 | 2.06 | 1.81 |
| Cu | 29 | fcc | 1 | 0.847 | 7.0 | 2.67 | 1.36 |
| Nb | 41 | bcc | 5 | 2.80 | 15.6 | 1.79 | 2.02 |
| Mo | 42 | bcc | 6 | 3.84 | 19.3 | 1.61 | 2.25 |
| Ag | 47 | fcc | 1 | 0.586 | 5.49 | 3.02 | 1.20 |
| W | 74 | bcc | 6 | 3.81 | 19.1 | 1.62 | 2.24 |
| Au | 79 | fcc | 1 | 0.590 | 5.53 | 3.01 | 1.21 |
| Pb | 82 | fcc | 4 | 1.32 | 9.47 | 2.30 | 1.58 |

The concentration of electrons in metals varies in the range $(0.25 - 5.03) \times 10^{29}$ m$^{-3}$. The Fermi energy ranges from 5.5 to 20 eV. The five and six-valency metals with the body-centered crystal lattice V, Cr, Nb, Mo, W have larger values of $\bar{n}$ than do metals with the face-centered crystal lattice Al, Ni, Cu, Ag, Pb, Au.

It is worth considering the correlation between the concentration of the electrons $\bar{n}$ in a metal and the parameters of strength of the crystal lattice.

We may choose a mean-square amplitude of the atom vibrations $\overline{u^2}$ as a measure of the interatomic bonds. The temperature dependence of this value $\overline{du^2}/dT$ is a characteristic of the crystal lattice strength of solid metals. The vibration of the atoms can be compared if one takes into account the different mass $m$ of the atoms. The value of $\overline{u^2}$ was found to be inversely proportional to the atomic mass [15]. Consequently, it is expedient to exclude the mass effect by multiplying values $\overline{du^2}/dT$ by the mass of one atom.

Thus, we have found [15] the reduced amplitude, that is, the parameter $\eta$, which characterizes the strength of the interatomic bonds,

$$\eta = m\frac{\overline{du^2}}{dT} \qquad (1.35)$$

The less the reduced amplitude $\eta$, the greater the strength of the crystal lattice.

In Figure 1.12 the correlation between the electron concentration and the parameter $\eta$ is presented.

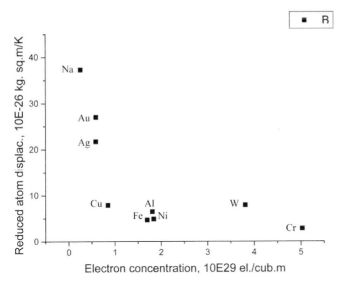

**Fig. 1.12** Correlation between the electron concentration in metals and the reduced amplitude of atomic vibrations.

The interatomic bonds in metals are realized by free electrons. From this point of view the correlation in Figure 1.12 seems to be logical. Also Figure 1.12 shows chromium as a material with strong interatomic bonds. Similar properties are indicated by the values of $\eta$ and $\bar{n}$ for tungsten, nickel, iron, and aluminum. It is no surprise that these metals are most favored in modern practice.

## 1.4.2
**Semi-Infinite Chain**

A simple model consists of a monoatomic linear chain fastened at one end. The end of the chain represents the surface. Assume that the axis $Oz$ is directed perpendicular to the surface, the point $z = 0$ is the abscissa on the surface.

The total potential of the electron is a periodic function of the distance. Thus, we can write

$$V(z + r) = V(z). \tag{1.36}$$

The dependence of the potential along the chain is assumed to be the cosine function,

$$V(z) = \begin{cases} 2V_0 \cos(2\pi z/a), & z < 0 \\ V_{vac}, & z \geq 0, \end{cases} \tag{1.37}$$

where $a$ is the parameter of the crystal lattice.

One tries then to solve the Schrödinger equation (1.25).

Far away from the surface, at $z \ll 0$, the electronic wave function in the lower-order approximation can be taken as a superposition of two plane waves:

$$\psi(z) = A \exp(ikz) + B \exp\left[i\left(k - \frac{2\pi}{a}\right)z\right] \tag{1.38}$$

where $A$ and $B$ are constants, $k = 2\pi/\lambda$ is the absolute value of the wave vector in the normal direction and $a$ is the crystal lattice parameter. Near the Brillouin zone boundaries characteristic band splitting occurs; an electron is scattered between states $k = +\pi/a$ and $k = -\pi/a$. The dependence of the energy of the electron on the wave vector is parabolic and splits near $k = \pm\pi/a$.

Let us briefly consider the solutions of the Schrödinger equation near a solid surface. The two solutions (for $z < 0$ and $z \geq 0$) have to coincide at the surface $z = 0$. The same requirement is imposed on the derivation $d\psi/dz$. It turns out that under the surface, the solution is a superposition of both an incoming and a reflected wave. Thus, a standing wave is formed near the surface.

In Figure 1.13 one can see the dependence of the squared wave function on distance. The peak at the surface indicates that the electron is attracted by the surface. At $z > 0$ the wave function decreases exponentially. The electronic tail is spread out the surface as far as 0.1 nm.

One can imagine the electron distribution near the surface as shown in Figure 1.14. A dipole layer is formed at the surface. The surface atoms are unbalanced because they have neighbors on one side and none on the other.

## 1.4 Distribution of Electrons near the Surface

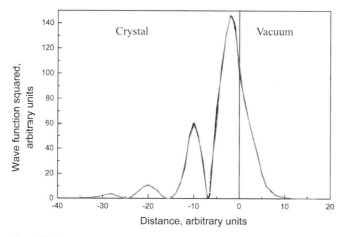

**Fig. 1.13** Squared wave function of an electron near the surface.

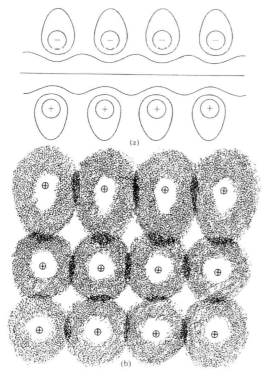

**Fig. 1.14** The distribution of electrons near the surface: a, equipotential lines in an electric double layer; b, distortion of the charge in a model simple cubic lattice. Reprinted from [2] with permission from Oxford University Press.

Therefore the electron distribution around them is an asymmetrical one with respect to the positive ion cores. This leads to a double electric layer as shown in Figure 1.14a. The dipoles are directed into the metal from the vacuum.

The graph in Figure 1.13 is a section of Figure 1.14b with a plane passing through ions parallel to the surface.

If the surface has steps the electronic charge tends to smooth them out (Figure 1.15).

**Fig. 1.15** Formation of electronic surface dipoles at the metal surface: a, smearing out of the charge distribution. Rectangles are the Wigner–Seitz cells; b, smearing out of the electronic charge distribution at a step. The increase in the negative charge (an electron cloud) is formed near the step. Reprinted from [4] with permission from Springer.

### 1.4.3
### Infinite Surface Barrier

One can further complicate the Sommerfeld model (perhaps more realistically).

The next step in the model of free electrons is to introduce a potential barrier at the surface. This barrier confines the electronic charge.

It is appropriate to replace the metallic cube by a slab of thickness $L$ in the $z$-direction which extends infinitely in the $x$- and $y$-directions [5]. One assumes that a discontinuity in the potential at the planes $z = 0$ and $z = L$ is described by

$$V(\vec{r}) = \begin{cases} 0, & \text{for } 0 \leq x \leq L, \quad -\infty \leq y, z \leq \infty, \\ \infty, & \text{elsewhere} \end{cases} \quad (1.39)$$

This means that one sets infinitely high potential walls at $x = 0$ and $x = L$.

Further, one replaces the actual infinite set of wave functions by a finite number, imposing boundary conditions, characterized by a period $L$ in the $y$ and $z$ directions and a fixed boundary at $x = L$.

In this case the appropriate wave functions have the form

$$\psi_k(x, y, z) = \left(\frac{2}{L^3}\right)^{1/2} \sin(k_x x) \exp[i(k_y y + k_z z)] \quad (1.40)$$

where $k_y = (2\pi n_y)/L$, $k_z = (2\pi n_z)/L$, $n_{y,z} = 0, \pm 1, \pm 2, \ldots$, $k_x = (\pi n_x)/L$; $n_x = 1, 2, 3 \ldots$

## 1.4 Distribution of Electrons near the Surface

The density of electrons is given by

$$\varrho_- = -|e| \sum_k |\psi_k|^2 \tag{1.41}$$

The variation in the density of the electronic charge can be expressed [1,5] as

$$\varrho_- = -n|e|B_r = -n|e|\left[1 + \frac{3\cos(2k_F z)}{(2k_F z)^2} - \frac{3\sin(2k_F z)}{(2k_F z)^3}\right] \tag{1.42}$$

where $z$ is the distance in the normal direction to the surface.

In Figure 1.16 the distribution of the electric charge near the surface is presented for two metals. The dependencies have been calculated in accordance with (1.42). The dependence exhibits some oscillations near the surface relative to the average bulk value. The wave length of the oscillations is equal to $\pi/2k_F$. The electron tails jut out of the metal surface for $\approx 0.08$ nm.

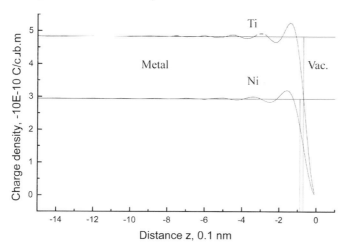

**Fig. 1.16** Distributions of the electric charge near the surface in nickel and titanium. The positive charge is assumed to be uniform inside the metal. The electron density oscillates in the metal near the surface. Dipoles are directed into the metal from the vacuum.

### 1.4.4
### The Jellium Model

The energy of two subsystems in metals determines the most characteristic features of electronic structure of a clean solid surface. This energy consists of the kinetic energy of the electrons, the energy of attraction between the electrons and the ions and the energy of electron repulsion.

Let us assume that a semi-crystal system consists of $N$ electrons and $N$ ions. The position of each ion is determined by the vector $\vec{R}$.

In quantum mechanics, the Hamiltonian is known to correspond to the total energy of a system. It is given by:

$$H = \sum_{i=1}^{N} \frac{p_i^2}{2m} + \sum_{\vec{R}} \sum_{i=1}^{N} \frac{Ze^2}{|\vec{r}_i - \vec{R}|} + \frac{1}{2} \sum_{i,j}^{N} \frac{e^2}{|\vec{r}_i - \vec{r}_j|} \qquad (1.43)$$

where $p_i$, $m$, and $e$ are the momentum, the mass and the charge of the electron, respectively, $Z$ is the charge of the nucleus and $\vec{r}_{i,j}$ are radius vectors of the electrons. The first term of (1.43) is the sum of the energies of the electrons. The second and third terms represent the Coulomb ion–electron and the electron–electron interactions. The presence of the last term makes the solution of (1.43) intractable.

The jellium model is a realistic approximation for simple metals. In this case the conduction electrons scatter only very weakly from the screened ion core pseudopotentials. In this model the positive charge is considered to be a homogenous medium forming a sort of jelly. The positive background charge has a density equal to the spatial average ion charge distribution. For the metal surface the distribution of the positive charge can be modeled by a step function:

$$n_+(z) = \begin{cases} \bar{n}, & z \leq 0 \\ 0, & z > 0 \end{cases} \qquad (1.44)$$

The axis $z$ is normal to the surface. The average electron concentration according to (1.31) is given by

$$\bar{n} = \frac{k_F^3}{3\pi^2} \qquad (1.45)$$

The uniform positive charge density $\bar{n}$ is often expressed in terms of an inverse sphere volume,

$$\frac{1}{\bar{n}} = \left(\frac{4\pi}{3}\right) r_s^3 \qquad (1.46)$$

The sum of the positive and negative charges must be equal to zero, so

$$\int [n(z) - n_+(z)] \, dz = 0,$$

where $n(z)$ is the electron density.

Replacement of the discrete lattice of ions and conducting electrons by the uniform positive background, leads to an error. However, the estimation shows that this error in energy is of the order of $10^{-7}$ eV which is much smaller than the Fermi energy.

The electron density variation perpendicular to the surface reveals two features. First, electrons jut out (spread out) into the vacuum ($z > 0$). In this way electrons create an electrostatic dipole layer at the surface. The electron distribution has no sharp edge. One should distinguish between the physical and the geometric surface. The first one denotes the location of the barrier, the second is the place where the positive charge background vanishes. The electron density decreases exponentially with the distance from the surface.

Second, the function $n(z)$ oscillates as it approaches an asymptotic value and compensates for the uniform positive charge in the bulk. The potential step keeps the electron within the crystal.

Figure 1.17 illustrates the dependence of the electron density upon the distance for two values $r_s$. At the higher electron densities ($r_s = 2$) the electrons extend further beyond the positive charge and the oscillations in the electron density are greatly diminished.

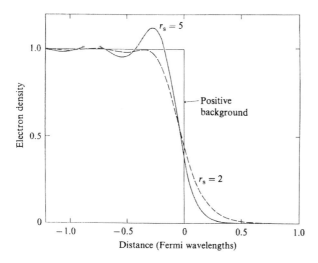

**Fig. 1.17** Electron density profile at a jellium surface. $r_s$ is the characteristic (Wigner–Seitz) radius. Data of Lang and Kohn. Reprinted from [3] with permission from Cambridge University Press.

We have already seen an analogous calculated dependence for nickel and titanium in Figure 1.16.

The distribution of the electron density near the monoatomic step is noteworthy. The electrons jut out as before but, in addition, they tend to smooth out the sharp step along the surface. Figure 1.18 presents the electrostatic potential near the monoatomic step for the jellium model. An electrostatic dipole appears oriented oppositely to the jut-out value. The net dipole moment is reduced relative to the flat-surface value.

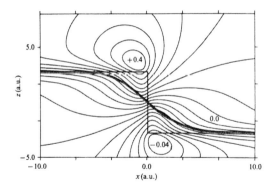

**Fig. 1.18** Electrostatic potential near a jellium step. The heavy solid curve indicates the smoothed electron surface. Data of Thomson and Huntington. Reprinted from [3] with permission from Cambridge University Press.

## 1.5
## Summary

The surface energy $\gamma$ is equal to the work necessary to form the unit area of surface by a process of cleavage of the solid into parts.

Variation in the Helmholtz free energy of a solid by the equilibrium cleaving process, in a general case, is given by

$$dF = -S\,dT - p\,dV + \mu\,dN + \gamma\,dA + A\,d\gamma \qquad (1.47)$$

The fourth term on the right-hand side describes the work needed to break down interatomic bonds and to create the surface area $dA$. The average area per atom is unchanged, but the number of atoms on the surfaces increases. The fifth term contributes the free energy due to changes in the interatomic distances on the new surfaces. These changes are balanced in the macroscopic and microscopic areas. The number of atoms on the surfaces remains unchanged. The stressed surface contributes a surface energy by an increment $d\gamma$.

A rearrangement of atoms occurs in the topmost layers of the surface. One uses a special notation to denote the surface superstructure. The notation reveals the dimensions, the nature and the orientation of the unit cell in the first plane relative to the underlying ones.

Defects always exist on real surfaces. Defects of zero dimension are vacancies, adatoms, ledge adatoms, and kinks. Their sizes in three dimensions are of the order of the interatomic distance. The step is a one-dimensional defect, in which the ledge separates two terraces from each other. Steps of the single atomic height prevail over steps of several atomic heights. There are also groups of defects such as divacancies, and steps of some interatomic distances in height. Other defects are related to dislocations. The applied stress and strain leads to the generation of ensembles of surface nanometric defects. These ensembles have the shape of prisms and differ from one another by dimension and energy of formation. Walls of nanodefects consist of steps of width from 5 to 50 nm. The walls are formed in consequence of an output of dislocations along planes of the light slip on the surface.

The parameters of the electron subsystem in some metals have been calculated based on the valency of the metals and the volumes of their crystal cells. The concentration of electrons in metals varies in the range $(0.25 - 5.03) \times 10^{29}$ m$^{-3}$. The Fermi energy ranges from 5.5 to 20 eV.

The electron concentration correlates with a parameter of the strength of the crystal lattice

$$\eta = m \frac{\overline{du^2}}{dT}$$

where $\overline{u^2}$ is the mean-squared amplitude of the atomic vibration, $m$ is the atom mass and $T$ is the temperature.

The free electrons in the metals are attracted by the surface. The electron density oscillates near the surface. At $z > 0$ the wave function squared decreases exponentially. The electronic tail juts (spreads) out on the surface as far as 0.08–0.10 nm. The electron distribution around the surface atoms is asymmetrical with respect to the positive ion cores. A dipole layer is formed at the surface. The dipoles are directed into the metal from the vacuum.

The jellium model is a realistic approximation for simple metals. In this case the conduction electrons scatter very weakly from screened ion core pseudopotentials. In this model the positive charge is considered to be a homogenous medium forming a sort of jelly. The positive background charge has a density equal to the spatial average ion charge distribution. For the metal surface the distribution of the positive charge can be modeled by a step function.

The electron density near the monoatomic surface step tends to smooth out the sharp step along the surface. An electrostatic dipole appears to be oriented opposite to the jut-out value.

# 2
# Some Experimental Techniques

The main aims of investigation of the strain surface are to determine the arrangement of atoms, to measure the residual stresses, to discover the properties of nanometric structures and to look for ways of strengthening the surface.

In this chapter we consider a number of techniques and particular experimental equipment.

## 2.1
## Diffraction Methods

An understanding of the phenomena at deformed surfaces on a microscopic scale is based specifically on the interaction between electromagnetic waves and matter.

The incoming electromagnetic wave interacts with atoms. The atoms excite and emit secondary waves, which interfere with each other. Thus, electron streams or X-rays are scattered by atoms in either a metal or an alloy to form diffraction patterns. These patterns provide information about the structural peculiarities of the subject under study.

### 2.1.1
### The Low-Energy Electron Diffraction Method

An electron beam is known to be also an electromagnetic wave.[1]

The elastic scattering of the electro-magnetic radiation, i.e. scattering without change of the wave length, can inform us about the lattice symmetry and about the arrangement of atoms. The inelastic scattering process allows one to estimate possible excitations of surface atoms.

The method should ensure the interaction of the irradiation with a thin surface layer. The initial beam must not penetrate too deeply into the material.

---

1) The wavelength $\lambda$ is given by the De Broglie equation, $\lambda = h/mv$, where $h$ is the Planck constant, $m$ is the mass of an electron, $v$ is the velocity of the electrons. The smaller is $\lambda$, the more effective the penetration of the radiation in a solid.

---

*Strained Metallic Surfaces.* Valim Levitin and Stephan Loskutov
Copyright © 2009 WILEY-VCH Verlag GmbH & Co. KGaA, Weinheim
ISBN: 978-3-527-32344-9

The nearly-ideal probes for the surface are low-energy electrons. They provide the most efficient technique. The depth of the penetration is extremely dependent on the energy of the electron waves. This dependence is presented in Figure 2.1.

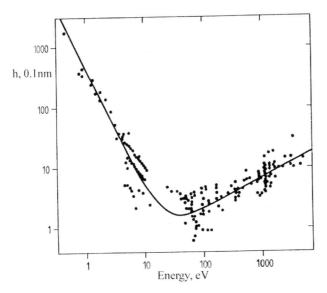

**Fig. 2.1** The mean free path of electrons versus their energy. Data for a large number of different materials [16].

The interplanar spacings in metals are of the order of 0.1–0.3 nm. One can see from Figure 2.1 that the free path of electrons with an energy of 10–100 eV has the same order. Consequently, the electrons with these energies can provide information about the surface layers. Electrons penetrate two or three atomic layers into the solid.

Figure 2.2 shows an experimental setup for the low-energy electron diffraction, that is for the so-called LEED technique. Electrons with energies in the range of 20–100 eV irradiate a crystal surface. The electrons, which scatter back elastically from the specimen surface, form diffraction patterns.

The apparatus consists of an electron gun, a system of electrostatic lenses, a step-up transformer, and a fluorescent screen or an analyzer of the electron energy.

The electron gun unit radiates the beam that is focused at the specimen surface by a lens system. The intensity of irradiation, which is diffracted by the specimen, is registered on the screen or is measured by the energy detector.

This beam scatters from the specimen surface through a set of charged grids, function of which is to catch electrons of lower energies. The grids are set

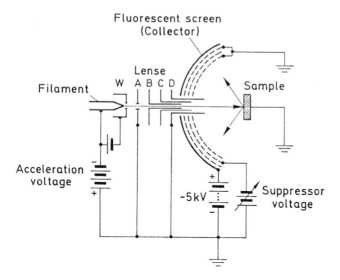

**Fig. 2.2** The scheme of an experimental installation for low-energy electron diffraction (LEED).

to retard all electrons other than those which have been scattered elastically. Thus, the grids select only the electrons which scatter from the surface elastically. The elastically scattered, diffracted electrons are then accelerated onto the fluorescent screen by applying about 5 kV. When they strike the screen, they cause the phosphor to glow, revealing a pattern of dots. In this way one obtains the diffraction pattern.

We know that all waves which are scattered by atoms interfere with each other. In order to find an interference function let us first consider waves that are scattered by two atoms.

The interference is shown in Figure 2.3. The first atom is situated in the origin $O$. The location of the second atom $P$ is determined by a vector $\vec{r}$. The direction of the incident beam is given by the wave vector $\vec{k}$, the direction of the scattered beam is determined by the vector $\vec{k}'$. Absolute values of the vectors are equal to each other: $|\vec{k}| = |\vec{k}'| = 2\pi/\lambda$.

The difference in the path lengths of scattered and incident rays is equal to $d = OB - AO$. One can obtain $d$ as the difference of the scalar products

$$d = (\vec{r} \cdot \vec{k}') - (\vec{r} \cdot \vec{k}) = (\vec{r} \cdot \vec{K}) \tag{2.1}$$

A wave displacement $Y$ at a point in space at time $t$ is equal to

$$Y = \frac{\Phi_0}{R} \exp i[(\omega t - (\vec{r} \cdot \vec{K})] \tag{2.2}$$

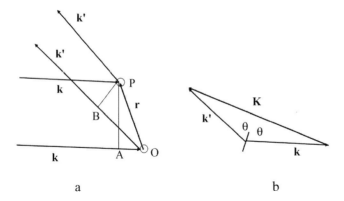

**Fig. 2.3** The scheme of scattering of electrons by two atoms (for details see text).

where $\Phi_0$ is the amplitude of the wave, $R$ is the distance from the origin $O$ to the point of observation. The wave displacements (2.2) should be summed up over all the crystal lattice points.

Let us consider a specimen of parallelepiped shape. The crystal lattice contains $N_a$ atoms along the side $\vec{a}$ of the elementary cell. Similarly, the crystal has $N_b$ nodes in the direction $\vec{b}$ and $N_c$ nodes in the direction $\vec{c}$.

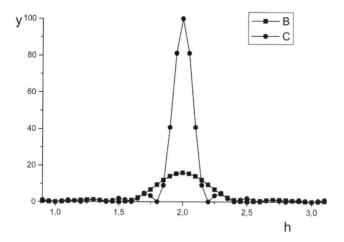

**Fig. 2.4** The interference functions: B, the number of reflected atomic layers is equal to 4, C, the number of atomic layers is 10. It is certain that the number of reflected layers affects the base width of the interference curve and, consequently, the length of the rods in the reciprocal lattice.

The intensity of diffracted irradiation at the distance $R$ from the crystal is given by [17]

$$I = \frac{|\Phi_0|^2}{R^2} \frac{\sin^2 N_a \Psi_a}{\sin^2 \Psi_a} \frac{\sin^2 N_b \Psi_b}{\sin^2 \Psi_b} \frac{\sin^2 N_c \Psi_c}{\sin^2 \Psi_c} \qquad (2.3)$$

where

$$\Psi_a = \frac{1}{2}(\vec{K} \cdot \vec{a}) = \frac{2\pi}{\lambda} a \sin\theta \cos\alpha$$

$$\Psi_b = \frac{1}{2}(\vec{K} \cdot \vec{b}) = \frac{2\pi}{\lambda} b \sin\theta \cos\beta$$

$$\Psi_c = \frac{1}{2}(\vec{K} \cdot \vec{c}) = \frac{2\pi}{\lambda} c \sin\theta \cos\gamma$$

$\alpha, \beta$, and $\gamma$ are angles between the vector $\vec{K}$ and directions $\vec{a}, \vec{b}, \vec{c}$, respectively. $2\theta$ is the angle between vectors $\vec{k}$ and $\vec{k}'$ (Figure 2.3b).

Denote

$$(\vec{K} \cdot \vec{a}) = \frac{2\pi a \sin\theta \cos\alpha}{\lambda} = \pi h$$

$$(\vec{K} \cdot \vec{b}) = \frac{2\pi b \sin\theta \cos\beta}{\lambda} = \pi k$$

$$(\vec{K} \cdot \vec{c}) = \frac{2\pi c \sin\theta \cos\gamma}{\lambda} = \pi l$$

where $h, k, l$ are certain numbers, which are not necessarily integers. Equation (2.3) may be rewritten in the general form as

$$I = \frac{|\Phi_0|^2}{R^2} \frac{\sin^2\left(\frac{\pi}{2} N_a h\right)}{\sin^2\left(\frac{\pi}{2} h\right)} \cdot \frac{\sin^2\left(\frac{\pi}{2} N_b k\right)}{\sin^2\left(\frac{\pi}{2} k\right)} \cdot \frac{\sin^2\left(\frac{\pi}{2} N_c l\right)}{\sin^2\left(\frac{\pi}{2} l\right)}. \qquad (2.4)$$

Let us examine only one of the multipliers of (2.4). This is a slice of the function (2.4):

$$y = \frac{\sin^2(\frac{\pi}{2} N h)}{\sin^2(\frac{\pi}{2} h)} \qquad (2.5)$$

Figure 2.4 presents the dependence of the diffraction pattern intensity on the number of reflected planes $N$ and the value of $h$. A crystal contains, in one dimension, only a few atomic planes. It is obvious that the interference peak is smeared if the number of reflected planes decreases.

It is convenient to examine data of electron reflections in terms of reciprocal space. The reciprocal lattice is known to be an image that is related uniquely

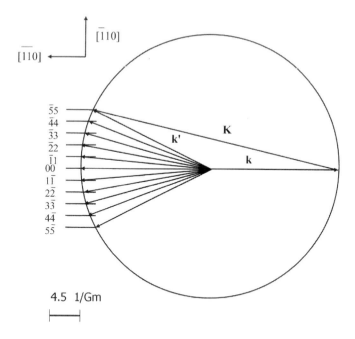

**Fig. 2.5** The scheme of scattering of the low-energy electrons in the reciprocal space of molybdenum. The energy of the electrons equals 20 kV, the wavelength is equal to $2.746 \times 10^{-10}$ m =0.2746 nm. The Ewald sphere of the radius $2\pi/\lambda$ is shown. $\vec{k}$ and $\vec{k}'$ are wave vectors of the incident and scattered waves, respectively. The vectors of the reciprocal lattice are $\vec{K} = \vec{k}' - \vec{k}$. The points of the reciprocal lattice are elongated to rods on account of the small thickness of the plate under examination. The rods are denoted as $00, \bar{1}1, \bar{2}2, \bar{3}3$ and so on. The length of the rods is given by $4/Na\sqrt{2} = 4.49$ Gm$^{-1}$. The distance between rods equals 2.25 Gm$^{-1}$. All values are shown on the same scale. Reflections occur in those directions $\vec{k}'$, where the Ewald sphere intersects the rods in the reciprocal lattice.

to the crystal lattice. The surface vectors of the reciprocal lattice are $\vec{a^*}$ and $\vec{b^*}$. For an orthogonal crystal lattice

$$a^* = 1/a; \ \vec{a^*} \| \vec{a}; \ b^* = 1/b; \ \vec{b^*} \| \vec{b}$$

The distance between adjacent points in the reciprocal lattice is inversely proportional to the distance between the points in the corresponding direction of the real lattice.

The position of a point $(hk)$ in the two-dimensional reciprocal lattice is determined by the vector $\vec{K}$:

$$\vec{K_{hk}} = h\vec{a^*} + k\vec{b^*}. \tag{2.6}$$

Vectors of the reciprocal lattice are of fundamental importance since they are equal to the difference of the diffracted and the initial wave vectors, see (2.1).

The scheme of the scattering of electrons by the (100) surface plane of molybdenum is presented in Figure 2.5. The scheme is drawn to scale. The incident beam is perpendicular to the plane (110) of the body-centered crystal lattice of molybdenum. Points of the crystal lattice are situated along the [$\bar{1}$10] direction at a distance of 0.449 nm. The free path of electrons is approximately equal to two atomic planes. The points of the reciprocal lattice are transformed into rods on account of a very few dimensions of the reflected area in ($\bar{1}\bar{1}$0) direction.

The formation of the electron pattern for the (110) plane of the molybdenum crystal lattice is illustrated in Figure 2.6c.

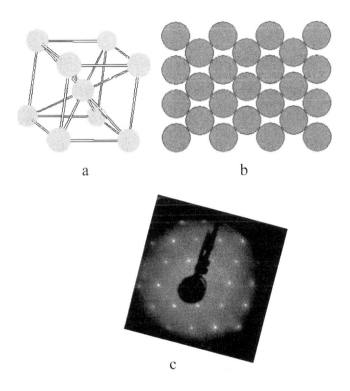

**Fig. 2.6** The electron diffraction of the (110) plane for a single crystal of molybdenum: a, the elementary cell of the body-centered crystal lattice; b, the arrangement of atoms in the plane (110); c, the diffraction image as the projection of the plane of the reciprocal lattice. The diffractogram c is turned with respect to b through an angle of $\pi/2$.

One can readily see that the LEED pattern is an image of the surface reciprocal lattice. The reciprocal lattice is directly observed in diffraction experiments. The surface reciprocal lattice is also a centered rectangular lattice. The LEED techniques have the ability to detect the two-dimensional symmetry of a few of the topmost atom layers.

The technique of low-energy electron diffraction is used to check the crystallographic quality of a surface and, especially, to determine the distortion of the unit cells near the surface. However, a detailed understanding of the intensity of LEED patterns involves the complex problem of describing processes of multiple scattering within the topmost atomic layers of the crystal. The atomic coordinates within the unit cell can only be obtained by measuring the intensity of the Bragg spots. For a LEED experiment a dynamic approach has to be applied to relate a structural model [4].

### 2.1.2
**The Reflection High-Energy Electron Diffraction Method**

The reflection high-energy electron diffraction (RHEED) technique operates in an energy range of 30–100 keV. The elastic scattering is strongly peaked in the forward direction with very little backscattering. Inelastic scattering mean-free-parts are relatively long (10–100 nm).

The electron beam is incident under grazing angles 3°–5° onto the surface of the specimen. This ensures that the electron penetrations into the surface remain relatively small.

The diffracted pattern is observed at similar angles on a fluorescent screen. The primary energy of the irradiation is much greater in the RHEED method than in the LEED method. The Ewald sphere diameter is much larger than the spacings in the reciprocal lattice. As a result diffraction patterns consisting of streaks correspond to the sections of rods. In fact a RHEED pattern is a planar cut through the reciprocal net.

RHEED has advantages in relative to LEED in relation multilayer surface processes. The use of grazing incidence angles makes the RHEED technique sensitive to the quality of the microscopic surface. The technique can identify changes in the surface morphology.

### 2.1.3
**The X-ray Measurement of Residual Stresses**

A residual stress is a tension or compression, which exists in the crystal lattice of a material without application of an external load.

Residual stresses are created as a result of mechanical operations which may be employed during part manufacture or by forces and thermal gradients imposed during use. The internal stresses are balanced within the volume of a specimen or of a machine element.

A variety of surface treatments is widely used for improving the characteristics of engineering components, particularly in the aircraft industry. Residual stresses can have a significant influence on the fatigue lives of the engineering crucial heavy-loaded parts.

Both the origination and growth of fatigue cracks are affected by the presence of residual surface stresses. The surface treatment must induce an appropriate distribution of residual stresses.

For the accurate assessment of fatigue lifetimes a detailed knowledge of the value of residual stresses and their distribution is required.

### 2.1.3.1 Foundation of the Method

The residual stresses in the metal surface layers can be classified as the macroscopic residual stresses and the microscopic residual stresses. X-ray diffraction provides a unique method to examine both types of stresses. The diffraction technique also allows the measurements to be done by a non-destructive method.

The macroscopic stresses (macrostresses) extend over distances that are large compared with the grain size of a metal or an alloy. They are balanced at distances of the order of the specimen dimension. The macroscopic stresses cause a uniform distortion of the crystal lattice and result in a shift of X-ray reflections (Figure 2.7a). One defines the macroscopic stresses as stresses of the first kind.

The microscopic stresses are placed in equilibrium over distances of the order of the grain size. Microstresses do not have a dominant direction and vary from point to point within the crystal lattice. The microscopic stresses result in a broadening of the X-ray diffraction peaks because of a set of $d$ values (Figure 2.7b). The microscopic stresses are defined as stresses of the second kind.

A scheme of the distribution of residual stresses in the crystal lattice is shown in Figure 2.8.

In Figure 2.9 the diagram of measurement of macroscopic residual stresses by the X-ray technique is presented.

The X-ray measurement of macroscopic residual stresses is based on the precise determination of strain $\varepsilon$ in the crystal lattice.

The Bragg formula is written as

$$n\lambda = 2d_{(hkl)} \sin\theta \tag{2.7}$$

where $n$ is an integer, $\lambda$ is the wave length, $d_{(hkl)}$ is the interplanar spacing of a family of atomic planes $(hkl)$ and $\theta$ is the Bragg angle.

From (2.7) one can obtain

$$\varepsilon = \frac{d - d_0}{d_0} = \frac{\Delta d}{d_0} = -\cot\theta \cdot \Delta\theta \tag{2.8}$$

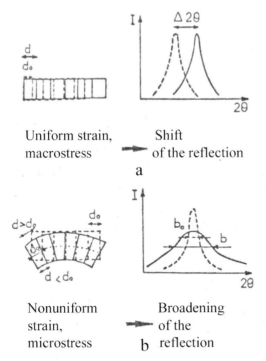

**Fig. 2.7** Influence of residual stresses on X-ray reflections: a, macroscopic stresses, the shift of a X-ray reflection occurs by reason of the uniform changing in the interplanar atomic spacing $d$, the shift equals $\Delta 2\theta$; b, microscopic stresses, the broadening of a reflection occurs because of the variation in the $d$ values, the broadening equals $b - b_0$ (after [18]).

Here $\varepsilon$ is the strain caused by the stresses, $d_0$ is the interplanar spacing in the unstrained crystal lattice, $d$ is the interplanar spacing in the strained lattice and $\Delta\theta$ is a shift in the X-ray reflection as a result of strain.

It is obvious that the closer $\theta$ is to 90° the larger shift of the reflection, since $\cot\theta$ approaches zero.

In order to find the deformation $\varepsilon$ it is sufficient to measure both the interplanar spacing in a specimen $d$ and in a non-strained standard $d_0$ made from the same material:

$$d = \frac{\lambda}{2\sin\theta} \tag{2.9}$$

$$d_0 = \frac{\lambda}{2\sin\theta_0} \tag{2.10}$$

The principal stresses $\sigma_1$ and $\sigma_2$ are considered as acting along the the surface and no stress is assumed to be normal to the surface, $\sigma_3 = 0$

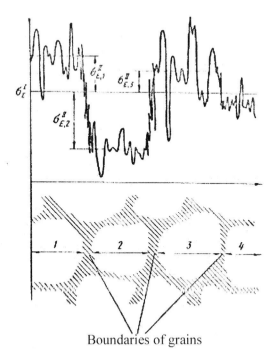

**Fig. 2.8** Distribution of residual stresses in the crystal lattice. Four crystallites 1–4 of a polycrystalline metal are shown. The macroscopic stress is $\sigma_\varepsilon^I$. The mean microscopic stress in the first grain is denoted as $\sigma_{\varepsilon,1}^{II}$, i.e., residual stress of the second kind in the grain number one. $\sigma_{\varepsilon,2}^{II}$ denotes the mean microscopic stress in the second grain and so on. The trend of the curves agrees with variable microscopic strains inside the crystallites.

The strain $\varepsilon_{\varphi\psi}$ in the direction defined by angles $\varphi$ and $\psi$ is given by [29]

$$\varepsilon_{\varphi\psi} = \varepsilon_x = \frac{(1+\nu)\sin^2\psi}{E}\left[\sigma_1\cos^2\varphi + \sigma_2\sin^2\varphi\right]\left[\frac{\nu}{E}(\sigma_1+\sigma_2)\right] \qquad (2.11)$$

It is possible, after some transformations, to express the stress $\sigma_x$ as

$$\sigma_x = \frac{E}{1+\nu}\frac{1}{d_0}\frac{\partial(d)}{\partial(\sin^2\psi)} \qquad (2.12)$$

where $E$ is the Young modulus, $\nu$ is the Poisson coefficient, $\psi$ is the angle between the perpendiculars to the specimen surface and to the reflecting crystalline plane $(hkl)$.

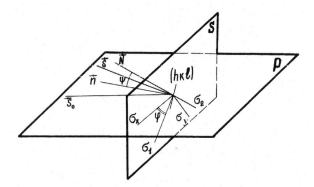

**Fig. 2.9** The measurement of residual internal stresses by the X-ray technique: $S$ is the surface of the specimen, $\vec{N}$ is the normal to the surface, $P$ is the equatorial plane, in which the X-ray beams transmit, $(hkl)$ is a reflected crystalline plane, $\vec{n}$ is the normal to the reflected plane, $\vec{s_0}$ is the initial beam, $\vec{s}$ is the reflected beam, $\psi$ is the angle between $\vec{N}$ and $\vec{n}$, $\sigma_1$ and $\sigma_2$ are the principal stresses, $\sigma_x$ is the stress in plane $P$ to be determined, $\varphi$ is the angle between $\sigma_x$ and $\sigma_1$, the stress $\sigma_y$ is normal to $\sigma_x$.

Equation (2.12) can be combined with (2.9) and (2.10) to obtain

$$\sigma_x = -\frac{E}{2(1+\nu)} \cot\theta \frac{\partial(2\theta)}{\partial(\sin^2\psi)} \tag{2.13}$$

The $2\theta - \sin^2\psi$ method based on the (2.13) has certain advantages. There is no need take the stress-free specimen in order to determine $d_0$. The method is non-contact, non-destructive and not time consuming.

One should record the X-ray reflections at several $\psi$ angles. Then it is necessary to plot a graph $2\theta - \sin^2\psi$ and calculate the derivative $\partial(2\theta)/\partial(\sin^2\psi)$. Further, one can calculate $\sigma_x$ in accordance with (2.13).

### 2.1.3.2 Experimental Installation and Precise Technique

In order to improve the accuracy and to increase the productivity of the X-ray technique we have made the following improvements to the standard procedure.

1. An increase in the operational angle $2\theta$.
2. The application of the $K_\beta$ X-ray irradiation and of a monochromator instead of using the usual $K_\alpha$-doublet.
3. The adjustment of the irradiated area on the surface under study by means of an optical system.
4. On-line data processing by means of a personal computer.

As the $2\theta$ angle approaches 180° one can appreciably increase the accuracy of measurements. However, the protective housing of the X-ray tube hinders

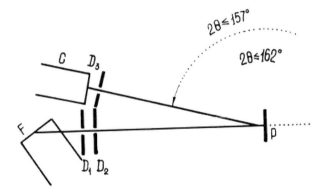

**Fig. 2.10** The standard scheme of the goniometer in the equatorial plane of a X-ray diffractometer. $F$ is the focus of the X-ray tube, $D_1$, $D_2$, and $D_3$ are slits, $P$ is the specimen, $C$ is the photoelectric counter. The maximum angle $2\theta$ is equal to $157°$ or $162°$ (without monochromator).

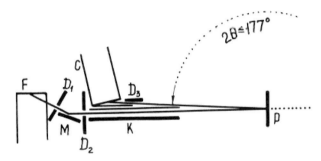

**Fig. 2.11** The improved scheme of the goniometer. The maximum angle $2\theta$ is increased to $177°$ (using a monochromator): $F$ is the focus of the X-ray tube, $M$ is the monochromator, $D_1$, $D_2$, and $D_3$ are slits, $K$ is the collimator, $P$ is the specimen, $C$ is the photoelectric counter.

the increase of the angle $2\theta$ [2]. Our aim was to eliminate this shortcoming. We have therefore worked out a new scheme for X-ray investigation.

We change the location of the scintillation detector and apply a new collimation system. The detector is turned in such a way that the plane of the detector window forms an acute angle with the scattered beam. The collimation system consists of a monochromator, slits and a narrow long collimator. The collimator has a very thin wall, which does not prevent the detector from moving and also protects it against the scattered irradiation. A single crys-

---

[2] Prevèy [19] presents a list of recommended diffraction techniques for various alloys. In this list the double Bragg angle $2\theta$ ranges only from 141 to $156°$.

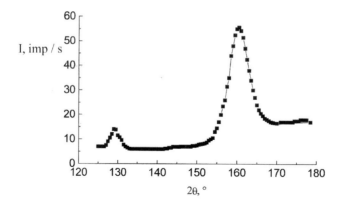

**Fig. 2.12** An example of X-ray diffractogram for a nickel-based superalloy: the reflections are (400) and (331). The $\beta$-radiation from the X-ray tube is produced with the cobalt anode, the wavelength $\lambda = 0.162075$ nm.

tal of graphite is applied as the monochromator. The irradiation area on the specimen surface is from 1 to 5 mm$^2$.

Figure 2.10 illustrates the standard scheme of the path of rays in the equatorial plane of the goniometer of a X-ray diffractometer. Our improved scheme for measurements of residual stresses is shown in Figure 2.11. The recorded Bragg angle $2\theta$ is increased up to $177°$. For the calibration of the diffractometer we use a reflection of $\alpha$-quartz at an angle of $174.5°$.

As a result, the accuracy of measurement of macroscopic residual stresses is increased by one order of magnitude.

Figure 2.12 presents an example of X-ray reflections for a superalloy. We record the same reflection $(hkl)$ from three to five times for every value of $\psi$. From two to five diffraction curves were taken for every fixed value of $\psi$. The obtained data are on-line processed by PC and are averaged. We have designed a program which allows one to perform the following operations:

- loading of the $I(2\theta)$ dependence that has been scanned under the given value of inclination of the angle $\psi$; the step and number of scanning are given in advance; points of experimental dependence are presented on the display;
- the variance analysis of the array in order to eliminate accidental surges: the modulus of the point deviation must be less than the dispersion;
- the plot of every diffraction peak and its smoothing, if it is necessary. The investigator may further choose an operation:
- the correction of the background position;
- the calculation of the center of gravity of the reflection, its integral intensity, the height of the peak and its width at half of its height;

- the plot of the $2\theta - \sin^2 \psi$ dependence and computation of the residual stress value according to (2.13);
- Fourier analysis of the $I(2\theta)$ reflection, calculation of the Fourier coefficients and comparison of the verifying synthesized curve with the experimental one.

## 2.1.4
**Calculation of Microscopic Stresses**

Fourier analysis of X-ray line broadening is used in order to measure microscopic stresses. As early as 1959 Warren reported this technique in detail for deformed metals [20].

This kind of reflection profile analysis is known as the Warren–Averbach method. One carefully records reflections of multiple orders. Apart from the specimen, one should study an annealed standard that contains no stresses or small subgrains.

The interference curves for a specimen and for the standard are expanded into the harmonic series. The Fourier coefficients enable one to find the structure parameters.

We apply Fourier analysis profiles to low and high-angle X-ray reflections. An annealed specimen of the same material is used as a standard. Every recorded curve is divided in 50 intervals and is expanded into a Fourier series. Fourier coefficients are calculated. The internal microscopic stresses $\varepsilon = \Delta a/a$ and the mean subgrain size $D$ are determined. The use of the PC allows us to reduce significantly the cumbersome calculations.

The dislocation density $\rho$ is calculated from these data in accordance with [21]:

$$\rho = \frac{4\pi^2 \varepsilon^2}{b^2 n} \tag{2.14}$$

where $b$ is the Burgers vector and $n$ is the number of dislocations in the pile-up.

## 2.2
**Distribution of Residual Stresses in Depth**

An etching technique is applied layer-by-layer for measurement of distribution of residual stresses in depth.

Specimens with a prismatic shape of dimensions $60 \times 20 \times 2$ and of the same alloy are used. The specimens are cut by means of an erosion machine to prevent hardening. These plates are placed in parallel to the treated material, therefore receiving the same treatment and strengthening. The treatment induces residual stresses in plates that become strained. The value of the deflection is appropriate for the measurement.

One performs electropolishing of superalloys in a water solution that consists of 2% hydrofluoric acid and 10% nitric acid.

The residual stress is expressed as

$$\sigma = \sigma_2 + \sigma_3 \qquad (2.15)$$

where $\sigma_2$ and $\sigma_3$ are stresses that arise as a result of the removing the external and the previous layer, respectively.

The stress $\sigma_2$ is given by

$$\sigma_2 = \frac{E(h-a)}{3L^2} \cdot \frac{\partial f}{\partial a} \qquad (2.16)$$

where $E$ is Young's modulus, $h$ is the height of the specimen, $a$ is the depth of the removed layer, $l$ is the length of the specimen, $\partial f / \partial a$ is the derivative of the bending deflection at a depth $a$.

One obtains the expression for $\sigma_3$ of the form

$$\sigma_3 = \frac{E}{3L^2}\left[4(h-a)f(a) - 2\int_0^a f(\xi)d\xi\right] \qquad (2.17)$$

where $f(\xi)$ is a function of the bending deflection of the specimen.

## 2.3
## The Electronic Work Function

The electronic work function is the difference between the energy of an electron outside a body (where forces acting on the electron are negligible) and the energy of the electron on the Fermi level. In other words, the work function is the smallest amount of energy required to remove an electron from the surface to a point just outside the metal with zero kinetic energy.

Outside the surface of a solid an electron has the electrostatic potential $\Phi_0$. The work function $\varphi$ is determined as

$$\varphi = -\Phi_0 - \frac{\mu}{e} \qquad (2.18)$$

where $\mu$ is the chemical potential, see (1.4), $e$ is the charge of the electron. $\varphi$ is a negative value, as the work is spent on the removal of the electron from the solid[3].

---

[3] From the many-electron point of view transferring an electron from the crystal to the point outside the crystal is an approximation. The definitions of the work function are considered in [5].

The measurements of changes in the work function enable one to determine the surface potential. The method of measurement of the work function is based on the determination of the contact potential difference.

Figure 2.13a illustrates the energy diagram for electrons in two metallic plates. Various metals have different Fermi levels. A difference in the levels is denoted in the Figure as $\Delta\phi$. When the two metal plates are connected electrically (Figure 2.13b) the Fermi energy must be uniform throughout the thermodynamic system. The same applies to the chemical potential of the electrons. The flow of electrons occurs from a metal with higher Fermi level to a metal with a lower one. As a consequence an electric double layer is formed, since the electron flow leaves plate 2 positively charged and plate 1 negatively charged. This leads to a discontinuous jump of the electrostatic potential at the junction, which compensates for the difference in chemical potentials. The difference in two potentials is just the difference between the respective work functions of the metals:

$$-e(\varphi_1 - \varphi_2) = V_1 - V_2 = -(\mu_1 - \mu_2), \tag{2.19}$$

where $V_1 - V_2$ is the contact potential difference and $-(\mu_1 - \mu_2)$ is the difference in the chemical potentials.

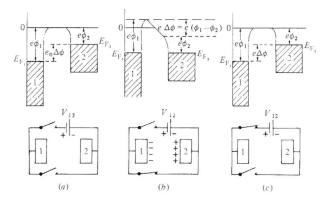

**Fig. 2.13** The contact potential difference in two metal plates: a, plates are isolated and charge free, $E_{F_1}$ and $E_{F_2}$ are the Fermi levels, respectively; b, plates are connected, electrons flow from plate 2 to plate 1, the Fermi levels are equalized; c, the balancing external potential is applied between the plates. Reprinted from [22] with permission from Cambridge University Press.

If an external potential is introduced in order to re-establish the corresponding vacuum levels, this potential will be equal, but opposite to, the work function difference. This is shown in Figure 2.13c.

In the well-known Kelvin technique the capacitor plates are caused to vibrate with respect to each other while the potential between the plates is measured. If no charge resides on the plates the potential is zero and remains zero as the capacity is altered. If, however, there is a charge on the plates, as a consequence of the contact potentials, then varying the capacitance generates a varying voltage between the capacitor plates. If an external potential is introduced in series with the capacitor and adjusted so that it is equal in magnitude and opposite in sign to the contact potential, then the net charge on the plates becomes zero and no change in potential will occur when the plates are vibrated. Thus, the contact potential difference is equal but opposite in sign to the adjusted voltage.

Lord Kelvin was the first to use the vibrating capacitor technique as early as 1898. He measured the charge manually using a quadrant electrometer (Figure 2.14). Later the Kelvin probe method was developed to produce many elegant versions.

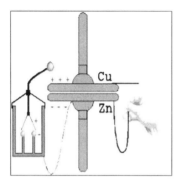

**Fig. 2.14** The set-up worked out by Lord Kelvin in 1898.

It is generally agreed that the electronic work function is one of the fundamental characteristics of a solid surface. The measurement of changes in the work function is used in the study. One should distinguish the work function data obtained from measurement on polycrystalline specimens from the work function of single crystal planes $(hkl)$.

## 2.3.1
**Experimental Installation**

An experimental installation for direct measurements of the electronic work function on the surface of metallic specimens, and also on the surface of gas-turbine blades and discs must meet the following requirements.
1. The absolute error in the work function determination should be less than 1 meV.
2. The measurement procedure should be computer-aided.
3. The installation must to enable one to carry out fatigue tests on specimens.

4. The installation must allow one to remove adsorbed atoms, at least partially, from the surface during investigation.

The accuracy of measurement is affected substantially by the noise level in the signal of the dynamic capacitor. The conventional technique of the Kelvin probe has a shortcoming: the signal-to-noise ratio decreases near the point of balance. We have attempted to achieve the signal–noise enhancement.

Figure 2.15 presents the experimental installation designed by us according to these requirements [23]. It is intended for measurement of the electron work function on the strained surface of metallic specimens with possible simultaneous fatigue tests.

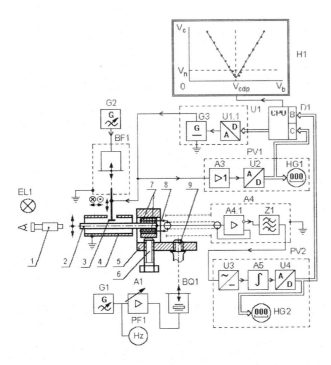

**Fig. 2.15** The experimental installation which enables one to measure the work function of specimens during fatigue tests. See text for details.

The installation includes the following basic units: a vibration table, a specimen assembly, an operating computer, a complex for balancing the potential, a measurement instrument, and a source of ultra-violet irradiation.

The specimen assembly is placed on a leg 9 of the piezoelectric vibration table BQ1 with the aid of a clamp 5.

The specimen 2 is fastened with a screw 6 between two pairs of isolating plates 7. A conductive foil 8 between the plates of each pair serves as an equipotential. A shield 4 is connected to earth in order to screen the specimen from externally induced electric fields.

The surface of the specimen 2 forms a capacitor with a vibrating scanning electrode 3. The scanning electrode (the probe) 3 has a cylindrical shape with a diameter of 1.40 mm and is made of gold.

The circuit used for producing the vibration of the specimen contains a generator of electrical oscillations G1, the amplifier A1, and the piezoelectric generator of mechanical oscillations BQ1. The frequency of the oscillations during fatigue tests is measured by the frequency indicator PF1. The microscope 1 serves for determine the amplitude.

The plane specimens (see below Figure 2.18) and resonance frequencies of the specimen oscillations are used.

Vibrations of the golden probe are generated by the generator of electric impulses G2 and by the electromagnetic vibration oscillator BF1. Micrometer screws enable one to adjust the probe along three coordinates and to scan the surface of the specimen with a step of 100 µm.

A balancing potential is applied to the probe due to the source G3 and the analog–digital transformer U1.1. The personal computer D1 controls the value of the balancing potential, which is measured by means of the amplifier A3, the analog–digital transformer U2 and the indicator HG1.

The circuit for measuring of the contact potential difference consists of the electrometric amplifier A4.1, the band-pass filter Z1, the digital voltmeter of alternating current PV2, and the indicator HG2.

The source of ultra-violet irradiation EL1 ensures refinement of the specimen surface from absorbed atoms.

### 2.3.2
**Measurement Procedure**

The specimen or the blade is placed between the holders. One cleans the surface under investigation using a solvent. The surface is irradiated by ultra-violet rays in order to remove of adsorbates from it.

The work function measurement involves approaching the probe to the surface at every point and then balancing the signal. The probe oscillates with an amplitude of 100 µm and with a frequency of 500 Hz. The minimum distance between the probe and the surface under study is of the order of 3–5 µm.

The balancing potential is adjusted by a computer program. One can see the potential curve on the screen of the indicator HG1.

The variable unbalanced signal arrives at the input of the amplifier A4. The amplifier contains an electrometric cascade A4.1 and the band-pass filter Z1. The filter has steps of 370, 520, 620 and 720 Hz and the band pass is 20 Hz.

This filter removes surface charge oscillations. The signal passes from the filter output to the digital voltmeter PV2. It is averaged, integrated and transformed into a digital code. Then the signal proceeds to the input of the computer D1 and it is seen on the indicator HG2 screen.

The computer draws the dependence of the average potential $V_c$ on the balancing potential $V_b$. This dependence is represented by two line segments (see Figure 2.15, top). The least-squares adjusted graph allows one to find the point of intersection $V_{cdp}$ at the abscissa. This point is the required contact potential difference.

The process of measurement is repeated in the course of scanning at every point on the surface.

The electronic work function of the specimen $\varphi_{spec}$ is calculated based on the known value of the work function of a standard $\varphi_{stand}$:

$$\varphi_{spec} = \varphi_{stand} - V_{cpd}, \tag{2.20}$$

where $V_{cpd}$ is the measured contact potential difference between the metal of the specimen and that of gold.

Since we use gold as the standard

$$\varphi_{spec}(eV) = 4.300 - V_{cpd}(eV) \tag{2.21}$$

The absolute error of the work function measurement became $\pm 1$ meV as a consequence of the improvement in the technique.

## 2.4
### Indentation of Surface. Contact Electrical Resistance

We have designed an experimental installation for the mechanical indentation of the surface layers and for measurements of the contact electrical resistance. The contact pair is a test spherical indenter and the surface of the specimen under examination.

The installation enables us to load the contact pair and to record the contact electrical resistance and the electromotive force continuously. Application of an impulse load or a cyclic loading is also possible.

The diameter of the indentor ranges from 5.0 to 22.5 mm. The installation makes it possible to test metallic pairs with a load of up to 3000 N with a rate of the static loading in the interval from 0.4 to 22.0 µm s$^{-1}$. A cyclic load may vary from 100 to 1500 Hz. Measurement of the electrical resistance in the interval from $10^{-7}$ to $10^6$ Ω is possible. The electrical resistance of the contact pair is measured with an accuracy of (0.01–0.50)%.

The installation allows the determination of the electromotive force. Figure 2.16 shows the scheme of the installation.

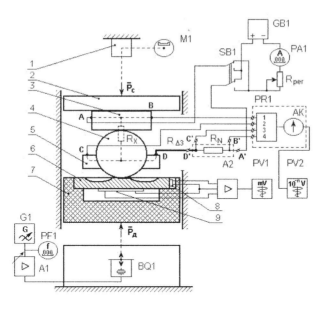

**Fig. 2.16** The experimental installation for measurement of the contact electrical resistance (see text for details).

The loading of the contact pair is performed with an electric motor 1 and a reducer M1. A unit of the oscillating load consists of a generator G1, an amplifier A1, and a piezoelectric vibration table BQ1. An indicator PF1 measures the load frequency. A specimen 3 is fastened between an supporting table 2 and a spherical indentor 4 with a washer 5. The washer 5 is used for attachment of electrical conductors. The indentor of the bearing steel is used. The vibrational load is applied to the specimen by an isolating cup 7 and a springing element 6. A straingage indicator 8 is glued on the element 6. The signal from straingage indicators through an amplifier A2 is recorded by a voltmeter PV1. Another indicator 9 compensates the thermoelectromotive force. It is glued on the unstrained surface of the springing element.

An electric balance is used to measure the contact electrical resistance. The measuring device involves an unknown resistance $R_x$, a standard resistance, $R_N$, a cross-connection $R_{\Delta 3}$, and a comparison circuit. An uncompensated signal is measured by a galvanometer AK and is recorded by an apparatus PV2. Points A–D and A'–D' correspond to balance points.

GB1 is a current source. The current is measured by an ammeter PA1. One can change the current direction through the circuit by switch SB1.

The external temperature is maintained constant with an error of $\pm 0.5°C$. Changes in the thermoelectromotive force are therefore related to the contact processes. The thermoelectromotive force can be expressed as

$$E = I(R_+ - R_-) \tag{2.22}$$

where $I$ is the current, $R_+$ and $R_-$ are the resistances of the pair for alternate directions of the current, respectively. The direction of the electric current is periodically alternated. In Figure 2.17 the sectional continuity of the output data is shown.

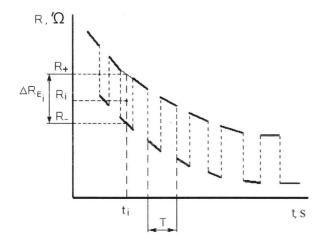

**Fig. 2.17** The contact resistance versus time and the direction of the electric current. T is the duration of the switching cycle.

The installation is intended for
- the study of the structure of contact surfaces;
- the determination of the actual contact area;
- the investigation of the effect of the surface relief on the contact stress.

## 2.5
## Materials under Investigation

The materials under investigation were pure metals, that is, aluminum, copper, and nickel; iron- and nickel-based superalloys; titanium-based alloys.

Vacuum-melted metals of 99.99% purity were examined. Alloys were of commercial purity.

Blades and discs of gas-turbine engines used in the industrial production were also studied in our investigation.

In Table 2.1 the compositions of the superalloys under study are shown. The superalloys were produced industrially. A standard heat treatment of every superalloy included solution treatment and a one-step or two-step ageing followed by air cooling.

**Table 2.1** Chemical composition (wt.%) of the superalloys under study.

| Alloy | Cr | Al | Ti | Mo | W | Co | Others | Fe | Ni |
|---|---|---|---|---|---|---|---|---|---|
| ZhS6K | 5.0 | 5.8 | – | 4.0 | 5.1 | 4.8 | – | – | rest |
| EI698 | 14.9 | 1.7 | 2.7 | 3.0 | – | – | 2.0 Nb | – | rest |
| EP866 | 16.0 | – | – | 1.5 | 0.7 | 5.0 | 0.3 Nb, 0.2 V | rest | 2.0 |
| EP479 | 15.8 | – | – | 1.0 | – | – | 0.07 N | rest | 2.1 |

The compositions of industrial titanium-based alloys are presented in Table 2.2.

**Table 2.2** Chemical composition (wt.%) of the titanium alloys under investigation.

| Alloy | Al | Mo | Cr | Si | Fe | Zr | Ti |
|---|---|---|---|---|---|---|---|
| VT 3-1 | 7.1 | 2.9 | 2.0 | 0.4 | 0.7 | – | rest |
| VT 8 | 6.5 | 3.3 | – | 0.3 | – | – | rest |
| VT 8M | 5.5 | 4.0 | – | 0.25 | 0.3 | 0.3 | rest |
| VT 9 | 7.0 | 3.8 | – | 0.35 | – | 2 | rest |

Specimens that were used for fatigue tests are shown in Figure 2.18.

## 2.6
## Summary

We have considered some effective techniques for investigating the strained surface layers of metals and industrial alloys.

One should distinguish between the macroscopic residual stresses and the microscopic residual stresses.

We apply the following procedures in order to improve the accuracy and to increase the productivity of the X-ray method of measurement of residual macroscopic stresses: an increase in the operational angle $2\theta$ up to $176°$; the use of the $K_\beta$ X-ray irradiation with a monochromator; the adjustment of the initial beam on the surface under study by means of an optical system; on-line data processing using a personal computer.

Fourier analysis of the X-reflection broadening is used to calculate the microscopic surface stresses.

The etching layer-by-layer technique is applied to determine the residual stress distribution in depth.

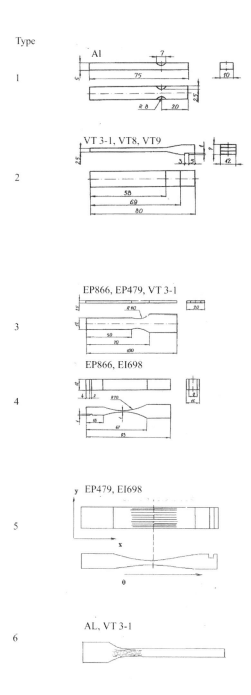

**Fig. 2.18** Specimens for fatigue tests.

The electronic work function is the minimum energy required to remove an electron from the surface to a point just outside the metal. In the well known Kelvin technique the capacitor plates are caused to vibrate with respect to each other whilst the potential between the plates is measured. If an external potential is adjusted so that it is equal in magnitude and opposite in sign to the contact potential, then the net charge on the plates becomes zero and no change in potential will occur when the plates are vibrated. Thus, the contact potential difference is equal but opposite in sign to the adjusted voltage.

We have designed an experimental installation, which enables us to measure the electronic work function on the strained surface of metal specimens and simultaneously to test specimens for fatigue. The installation has important advantages: the signal-to-noise ratio near the point of balance is increased; the absolute error in the work function determination is as low as $\pm 1$ meV; the process of measuring is computer-aided; one can remove adsorbed atoms from the metallic surface and it is possible to test specimens or blades for fatigue.

We have worked out the experimental installation for the mechanical indention of the surface layers and for measurement of the contact electrical resistance. The contact pair consists of a test spherical indenter and the specimen surface to be studied. The set-up enables us to load the contact pair and to measure and to record the contact electrical resistance and the electromotive force continuously. It is also possible to apply an impulse and a cyclic loading.

Strained surfaces of pure metals, nickel- and iron-based superalloys, titanium-based alloys, as well as that of gas-turbine blades and discs have been studied.

# 3
# Experimental Data on the Work Function of Strained Surfaces

In this chapter we consider the effect of strain on the electronic work function of metals and alloys.

## 3.1
## Effect of Elastic Strain

Studies of the electronic work function for elastic deformed metals are quite scarce. This is no surprise since the elastic strain limit of soft metals is very low. Initially, the elastic deformation under a low applied stress has an essential plastic component as soon as one increases the stress. It occurs before one can record the effect of the pure elasticity on the work function. According to our data, the work function increases by tenths of a millivolt as a result of elastic tension (Table 3.1).

**Table 3.1** The effect of the tensile elastic strain $\varepsilon$ on the increment of the work function $\Delta\varphi$, meV.

| $\varepsilon$ | 0.01 | 0.02 | 0.03 |
|---|---|---|---|
| Al | +17 | +35 | +40 |
| Ni | +14 | +24 | – |

The authors [24] investigated both elastic and plastic areas of strain. Beam-shaped specimens of dimensions $3 \times 3 \times 140$ mm$^3$, were polished to obtain a mirror-finished surface. The deformation was introduced by bending the beam specimen, so that one can obtain both a tensile and a compressive strain with different magnitudes in different positions of the specimen cross-section. The strain at each point was calculated by the authors taking into account the degree of bending.

The effect of elastic deformation on the electronic work function is presented in Figure 3.1 for aluminum, and in Figure 3.2 for copper. One can see that the work function increases under the elastic strain of 0.003 by 30 and 58 meV for aluminum and copper, respectively. However, according to the authors' data this effect takes place for both metals only if the deformation

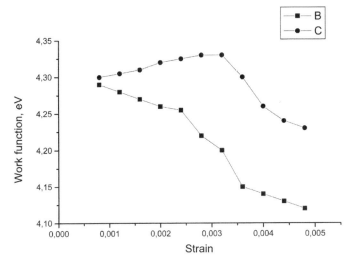

**Fig. 3.1** The electronic work function of aluminum versus the strain: B is the tension; C is the compression. Experimental data on bending tests from [24].

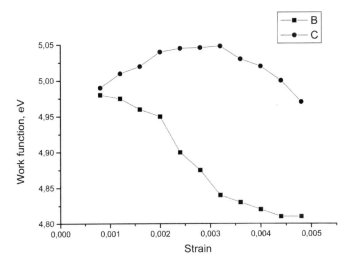

**Fig. 3.2** The effect of strain on the electronic work function of copper. B is the tension; C is the compression. Experimental data on bending tests from Ref. [24].

is compressive (curves C). The position of the maximum depends weakly on the metal. The drop of curves after the maximum appears to be because of transition from the elastic strain to the plastic strain.

## 3.2
## Effect of Plastic Strain

Figure 3.3 shows the results of our investigation into the influence of plastic strain on the work function for aluminum. Double shovel shaped plane specimens with a gauge width of 10 mm and the gauge length up to 50 mm were used.

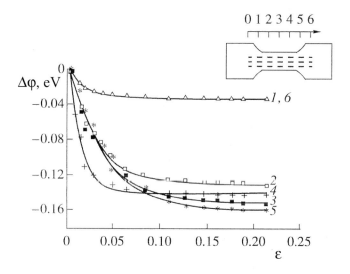

**Fig. 3.3** Three lines of the scan along the aluminum specimen with six segments are shown in the upper part of the Figure. The dependence of the work function on the plastic strain for different segments of the specimen: curves 1 and 6 correspond to segments 1 and 6, respectively; curve 2 corresponds to segment 2; curve 3 corresponds to segment 3; 4 to segment 4; 5 to segment 5. The crack was initiated at the segment 5. Symbols represent experimental data, solid curves were calculated in accordance with (3.1).

One can see that the work function decreases significantly when the strain increases. The change in the work function on the gauge in comparison with the unstrained material is equal to $-160$ meV.

An increment in the work function is calculated using an interpolation formula. A suitable empirical relationship appears to be

$$\Delta\varphi = \varphi_0\{1 - \exp[-\alpha(\varepsilon - \varepsilon_0)]\} \tag{3.1}$$

where $\varphi_0$ is an initial value of the work function, $\alpha$ is a dimensionless parameter of the deformation, $\varepsilon$ is the strain, $\varepsilon_0$ is the strain that corresponds to the commencement of plastic deformation of the material.

Figure 3.4 presents the distribution of the work function on the surface at various strains. The transition to the plastic strain stage at $\varepsilon = 0.005$ causes a characteristic decrease in the work function along all lines of measurement.

The greater the strain in the gauge portion of the specimen, the greater decrease in the work function. The decrease in the work function is also observed for other metals during plastic strain (Figure 3.5).

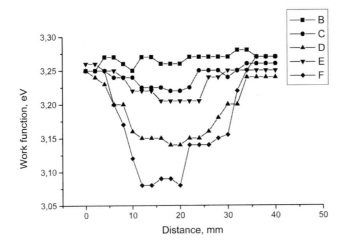

**Fig. 3.4** Distribution of the work function over the surface of the aluminum specimen: B, before loading, the strain $\varepsilon = 0$; C, $\varepsilon = 0.005$; D, $\varepsilon = 0.110$; E, $\varepsilon = 0.110$ and unloading for 24 hours; F, $\varepsilon = 0.215$. The specimen shape is shown in Figure 3.3.

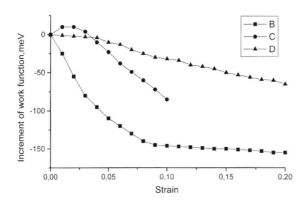

**Fig. 3.5** Effect of plastic strain on the electronic work function for some metals: B, aluminum; C, nickel; D, copper. B and C, the tension, experimental data of the authors; D, the compression, data from [25].

We have studied the dependence of the work function on the strain under loading and unloading of specimens. The experiment proceeds as follows:
- The loading of the specimen ⇒ the strain; the unloading.

- Again the loading ⇒ the strain; the unloading.
- Again the loading and so on.

The typical data obtained are presented in Figure 3.6. A slight increase in the work function (up to 3–7 meV) is observed in the region of elastic deformation. Plastic deformation up to 0.0025 corresponds to a decrease in the electronic work function from 3.24 to 3.08 eV. Stopping the loading leads to an increase in the work function. Thus, one can see that the effect of the loading on the work function is, to a certain extent, reversible in the plastic area. Ascending branches at the strain curve (Figure 3.6, curve 2) are in accordance with descending branches of the work function curve (Figure 3.6, curve 1).

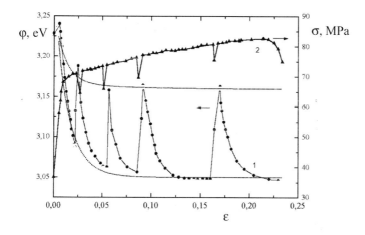

**Fig. 3.6** Dependence of the work function on strain: 1, the work function versus strain; 2, the applied stress versus strain. Solid lines correspond to areas of variation in the work function during strain and unloading. The same aluminum specimen was used for the measurement.

There is a minimum value of the work function. It is achieved at $\varepsilon = 0.24$ and equals 3.05 eV for plastic deformed aluminum.

The decrease in the work function during plastic strain also occurs in industrial superalloys. In Figure 3.7 the curves obtained are presented. The maximum variation in potential relief occurs at the segment of the largest strain.

We examined gas-turbine compressor blades of the VT3-1 titanium-based alloy. The length of the blades under investigation is 80 mm. Several blades were annealed at a temperature of 923 K for 3 hours. The strengthening of the blades is carried out by means of a special installation with ball-bearings as working tools. The balls are put in motion by the chamber walls vibrating in the ultrasonic field with a frequency of 17.2 kHz and an amplitude of 45 μm. The work function is measured along the backs and edges of the blades. Experimental points are located at intervals of 1 mm.

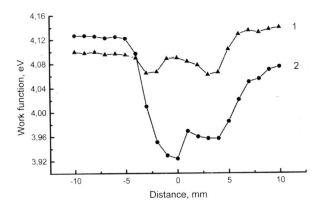

**Fig. 3.7** The distribution of the work function along strained specimens of superalloys. The tensile deformation of specimens was carried out up to the yield stress. Point 0 is the center of the specimen gauge: 1, the EP479 superalloy; 2, the EP866 superalloy.

There is a great difference between the annealed and strengthened blades (Figure 3.8). The severe surface deformation causes a drop in the work function down to −400 meV. The variation in the work function on the strengthened surfaces of the blades is about ±100 meV.

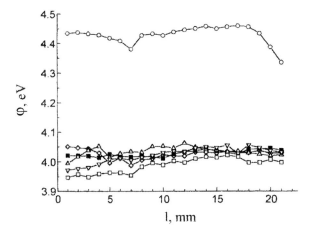

**Fig. 3.8** Distribution of the work function along the back of the blades: ○, data for an annealed blade; the other five data points correspond to five blades after surface treatment by ball-bearings in the ultrasonic field.

## 3.2.1
**Physical Mechanism**

Several physical mechanisms have been proposed to explain the experimental data. Some of them appear to be either speculative or based on a general understanding.

- The decrease in the work function is related to the formation of new surfaces as a result of deformation.
- The decrease results from an effect of the plastic strain on the Fermi level.
- The distortion of the crystal lattice also plays a role (as well as other factors).
- The decrease in the work function of plastic strained metals depends on the change in the surface roughness.
- An increase in the dislocation density results in the drop of the work function.
- The decrease in the work function is due to the formation of steps on the surface of strained metals. The surface steps are the consequence of the output of deforming dislocations.

We consider only this last situation since it is the only theory that is well-grounded in experiment.

The authors of [26] studied in detail the influence of steps on the work function for platinum, gold and tungsten single crystals. By spark erosion machining the surfaces were given cylindrical shapes in order to produce a locally varying step density in one direction.

The flat tungsten surfaces, inclined at various degrees to (100) plane demonstrated the characteristic LEED features of stepped surfaces. All the investigated tungsten surfaces are formed by terraces of (100) orientation separated by monoatomic steps. The LEED patterns indicate that curved platinum and gold surfaces are formed by (111) terraces as well as being separated by monoatomic steps. The terrace width decreases continuously (increasing step density) with increasing inclination angle towards the (111) plane.

The effect of surface steps on the work function is illustrated in Figure 3.9. The step density varied across the surface between $1.0 \times 10^8$ and $7.0 \times 10^8$ m$^{-1}$.

We consider the evidence of Besocke, Krahl-Urban and Wagner [26] as crucial experiments. They have proved by direct experiment that the main cause of the decrease in the work function for plastic strained metals is really the formation of surface steps.

Let us look again at Figures 1.15 and 1.18. There is an electron cloud near the metal inside the right angle that is formed by the step. It is obvious that this cloud facilitates the removal of the electron from the metallic surface.

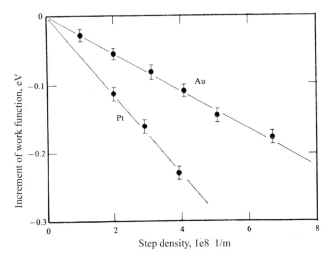

**Fig. 3.9** Dependence of the work function increment on the density of surface steps. Experimental data of [26] for gold and platinum. One can see that the work function decreases linearly with an increase in the step density on the surface.

The dependence of the work function on the density of monoatomic steps on the surface of a crystal is expressed as

$$\Delta\varphi = \frac{1}{\varepsilon_0} e P n \tag{3.2}$$

where $\varepsilon_0$ is the electric constant, $e$ is the electron charge, $P$ is the dipole moment per unit length of the surface step and $n$ is the density of surface steps.

Deforming dislocations emerge from the surface during strain and generate steps. The surface step has its own dipole moment. Estimation of the dipole density is possible according to (3.1) and (3.2). Results of estimation are presented in Figure 3.10. The density of dipoles is found to be of the order of $(6-9) \times 10^6$ m$^{-1}$ for 3–5% strained aluminum. The curves $n - \varepsilon$ are found to achieve a limit after the surface dipoles amount to a certain density.

Consequently, the work function of metals and alloys is appreciably affected by the strain. It is useful to study the work function for strained metallic surfaces. As early as 1990 [95] we proposed to term this effect *the strain-emission phenomenon*.

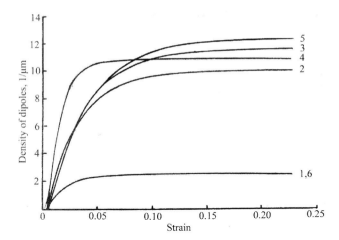

**Fig. 3.10** Density of dislocation dipoles versus strain. The numbers on the curves denote the different segments of the specimen, see Figure 3.3.

## 3.3
## Influence of Adsorption and Desorption

Ultraviolet irradiation of the surface of aluminum specimens results in a displacement of the curves $\varphi - x$ to lesser values (Figure 3.11, curve 2). The change in the work function is approximately $-200$ meV. This effect is assumed to be caused by clearing the surface and removal the adsorbed atoms. Adatoms absorb the electromagnetic irradiation, become excited, and cannot stay on the surface. Plastic strain leads to a further decrease in the work function (Figure 3.11, curve 3).

The effect of ultraviolet irradiation is found to be reversible. The initial value of the work function is restored if the irradiation is turned off. Figure 3.12 illustrates these events.

Desorption can also be caused by heating the specimen. The process of thermal desorption is also reversible. Figure 3.13 shows a decrease in the work function by $-450$ meV after heating by 200 K. So heating influences the work function similarly as the ultraviolet irradiation.

The authors [27] proposed that the effect of the adsorption on the electronic work function could be expressed as

$$e \cdot \Delta\varphi = \pm e \frac{p_0 n_a}{\varepsilon_0} \left(1 + \frac{\alpha \zeta \delta n_a^{\frac{3}{2}}}{4\pi\varepsilon_0}\right)^{-1} \tag{3.3}$$

where $e$ is the charge of the electron, $p_0$ is the dipole moment of an adsorbed

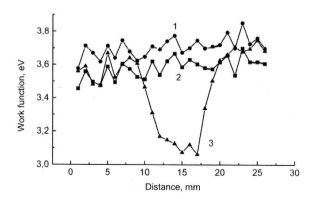

**Fig. 3.11** The potential relief of an aluminum specimen. 1, initial state of the specimen; 2, as 1 but with the ultraviolet irradiation; 3, as 2 after the 0.2 tension strain.

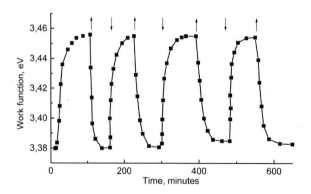

**Fig. 3.12** Variations in the electronic work function of the aluminum surface as a result of ultraviolet irradiation: ↑, the source of irradiation is turned on; ↓, the source is turned off.

atom, $n_a$ is the density of adsorbed atoms, $\alpha$ is the polarizability, $\zeta$ is a constant that is dependent on the interaction between surface dipoles, $\delta$ is a parameter that is dependent on the ordering of dipoles, and $\varepsilon_0$ is the electric constant [1].

1) For instance, for the H$_2$O molecule $p_0 = 3.33 \times 10^{-30}$ C·m, $\alpha = 1.44 \times 10^{-30}$ m$^3$ [28], $\zeta = 11.03$, $\delta = 1$ if the molecules are distributed uniformly.

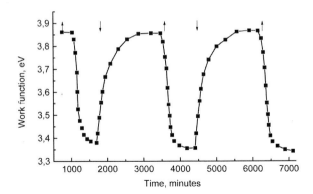

**Fig. 3.13** Variations in the electronic work function of the aluminum surface as a result of heating: ↑, heating to 493 K; ↓, cooling of the specimen to room temperature. The initial treatment of the specimen was annealing in vacuum at 523 K for two hours and the subsequent relaxing for 96 hours at 293 K.

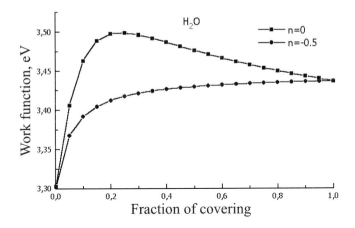

**Fig. 3.14** The work function of the aluminum surface versus the fraction of covering by adsorbed molecules of water. The curve $n = 0$ corresponds to the uniform distributions of adsorbed molecules, the curve $n = -0.5$ corresponds to a random distribution of adsorbed molecules.

$n_a = b\vartheta/d^2$, where $d$ is the parameter of the plane crystal lattice, $\vartheta$ is the fraction of the surface covered by adsorbed atoms $(0 \leq \vartheta \leq 1)$, $b = 1$ for the square cell and $b = 2/\sqrt{3}$ for the triangular cell. $\delta = 1$ if the distribution of adsorbed atoms is uniform and $\delta = \frac{1}{\sqrt{\vartheta}}$ if adsorbed atoms are distributed

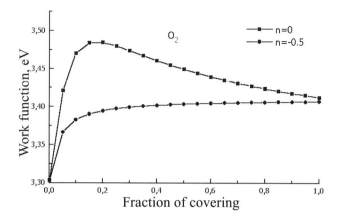

**Fig. 3.15** As Figure 3.14 for molecules of oxygen.

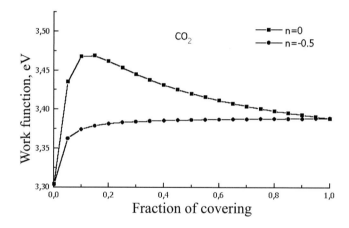

**Fig. 3.16** As Figures 3.14 and 3.15 for molecules of carbon gas.

randomly. Substituting the values into (3.3) we find

$$\Delta\varphi = \pm\frac{p_0 b\vartheta}{\varepsilon_0 d^2}\left[1 + \frac{p_0^2 \xi \vartheta^n (b\vartheta)^{\frac{3}{2}}}{12\pi\varepsilon_0^2 d^2 kT}\right]^{-1}. \tag{3.4}$$

Figures 3.14, 3.15 and 3.16 present results of our calculations according to (3.4). The work function increases sharply under the effect of adsorption. Its value is more affected by molecules of water, than by molecules of oxygen or carbon gas. The increment $\Delta\varphi$ is approximately +100 meV even if only 20% of the metallic surface is occupied by adsorbed $H_2O$ molecules.

## 3.4 Summary

There is little evidence for the influence of elastic strain upon the work function of metals and alloys. It appears that the work function increases by $+15-+50$ meV a result of elastic strain.

The transition from elastic strain to the region of plastic strain causes a characteristic decrease in the work function by tens or hundreds of meV.

The greater the strain in the gauge portion of specimens, the greater the decrease in the work function. The change in the work function in comparison with the unstrained material is of the order of $-160$ meV, $-50$ meV, and $-80$ meV for aluminum, copper, and nickel, respectively. The drop in the $\varphi$ value is found to be equal to $-400$ meV for the strained surface of gas-turbine blades.

In stress – strain and work function – strain graphs ascending branches of stress are in agreement with descending branches of the work function curves and vice versa.

The decrease in the work function is due to the formation of steps on the surface of strained metals. An electron cloud exists near the metal inside the right-angle that is formed by the step. This cloud facilitates the removal of the electron from the metallic surface. The work function decreases linearly with an increase in the step density on the surface. Surface steps are the consequence of the emerge of deforming dislocations on the surface during deformation.

The ultraviolet irradiation of the surface results in a displacement of the curves $\varphi - x$ to lesser values. This effect is assumed to be caused by clearing of the surface and removal adsorbed molecules. Adsorbed molecules which are excited by the electromagnetic irradiation cannot remain on the surface. The effect of the ultraviolet irradiation on the work function of the surface is revealed to be reversible as is the effect of heating. The work function increases sharply (by $+80-+200$ meV) under the effect of adsorption. Its value is more affected by molecules of water than by molecules of oxygen or of carbon dioxide.

# 4
# Modeling the Electronic Work Function

In the previous chapter we considered the experimental data on the effect of strain on the electronic work function. Recall that the work function increases weakly in the case of elastic deformation and decreases appreciably during plastic strain.

It is the object of this chapter to describe physical models of the work function for the strained metallic surfaces. We use the jellium model to investigate the effect of strain on the surface energy and the work function. In jellium model of metals the positive charge of ions is distributed uniformly over the volume (see subsection 1.4.4). Electrons are moving through the uniform, positively charged medium created by metallic ions. We further consider the development of the jellium model. We then modify this model for relaxation of the crystal structure and for a discontinuity of the ionic charge. We devote the final section to a model of neutral orbital electronegativity.

## 4.1
## Model of the Elastic Strained Single Crystal

The higher the density of atoms in a crystal plane the greater the value of the work function [29]. This evidence allows one to assume that the elastic strain of a single crystal results in a formation of a preferred direction. Consequently, an induced anisotropy is generated in the crystal lattice. This leads to a change in the atomic density of parallel planes and in the distance between the planes. The concentration of free electrons is also affected by the preferred direction.

Consider the elastic strain of a cubic face-centered single crystal along the [100] direction. The elastic deformation of the crystal is shown in Figure 4.1. The volume of the unit cell $V$ depends on the strain $\varepsilon$ as

$$V = V_0[1 - (1 - 2\nu)\varepsilon] + o(\varepsilon^2) \tag{4.1}$$

where $V_0$ is the volume of the unstrained unit cell, $\nu$ is the Poisson coefficient and the term $o(\varepsilon^2)$ represents infinitesimals of higher order. $V_0 = a_0^3$, where $a_0$ is the parameter of the unit cell. The volume electron concentration for

*Strained Metallic Surfaces.* Valim Levitin and Stephan Loskutov
Copyright © 2009 WILEY-VCH Verlag GmbH & Co. KGaA, Weinheim
ISBN: 978-3-527-32344-9

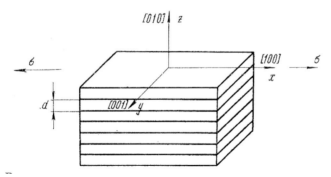

**Fig. 4.1** Scheme of the elastic strain of a single crystal. The stress $\sigma$ is applied along the $Ox$ [100] axis. The work function is measured along the $Oz$ [010] axis.

trivalent aluminum (four atoms in the unit cell) is given by

$$n_+ = \frac{12}{V}$$

One can express the total energy of the crystal as

$$E = E_j + E_M + E_{ps} \tag{4.2}$$

where $E_j$ is the energy of the electron gas in the jellium model, $E_M$ is the Madelung energy, $E_{ps}$ is the first order of energy of the electron – ion interaction in the expansion of the Ashcroft pseudo-potential.

We obtain expressions for the elastic strained crystal of the form [30]

$$a = a_0(1+\varepsilon) \tag{4.3}$$

$$d = \frac{a_0}{2}(1-\nu\varepsilon) \tag{4.4}$$

$$c = \frac{a_0}{\sqrt{2}}[1 + (0.5 - \nu)\varepsilon] \tag{4.5}$$

where $a$ is the distance between atoms in the [100] direction, $d$ is the distance between the (010) atom planes in the [010] direction and $c$ is the distance between atoms in these planes, that is inside planes, which are perpendicular to the $Oz$ axes.

We define the distribution of the electron density along the [010] axis by the one-parameter test functions as follows:

$$n_1(z) = \frac{n_+}{1 + \exp(b_1 z)} \tag{4.6}$$

or

$$n_2(z) = n_+ \begin{cases} 1 + \frac{1}{2}\exp(b_2 z), & z < 0 \\ \frac{1}{2}\exp(-b_2 z), & z > 0 \end{cases} \tag{4.7}$$

We choose two test functions in order to reveal the influence of the electron density distribution. One needs to find the parameters $b_1$ and $b_2$. We can find values of these parameters from a condition of the minimum of the surface energy:

$$\frac{d\gamma(b_i,\varepsilon)}{db_i} = 0 \qquad (4.8)$$

where $i = 1, 2$, and the surface energy $\gamma$ is dependent on $b_i$ and on the strain $\varepsilon$. The thickness of the electron layer outside of surface is approximately equal to the inverse values of $b_i$, that is $h \approx 1/b_i$; assuming the atomic system of units $e = \hbar = m = 1$.

The specific surface energy is also given by

$$\gamma = \gamma_j + \gamma_M + \gamma_{ps} \qquad (4.9)$$

We have obtained [30] expressions for three components of the surface energy of the form:

$$\gamma_j = I_1 \frac{n_+^2}{b^3} + \frac{1}{b}\left[-I_2 n_+^{\frac{5}{3}} + I_3 n_+^{\frac{4}{3}} + I_4 n_+\right] + I_5 n_+ b \qquad (4.10)$$

where the values of $I_{1-5}$ are given in Table 4.1.

**Table 4.1** Values of $I_j$ in (4.10).

|  | $I_1$ | $I_2$ | $I_3$ | $I_4$ | $I_5$ |
|---|---|---|---|---|---|
| $n_1(z)$ | 3.768 | 2.184 | 0.329 | $8.556 \times 10^{-3}$ | $6.940 \times 10^{-3}$ |
| $n_2(z)$ | 1.571 | 1.642 | 0.250 | $6.587 \times 10^{-3}$ | $9.630 \times 10^{-3}$ |

$$\gamma_{ps} = \sum_{k=1}^{\infty} \frac{(-1)^{k+1}}{k^3} I_6 \qquad (4.11)$$

$$I_6 = \frac{4\pi(n_+)^2}{b_1^3}\left[1 - \frac{kx \cosh(ky)\exp(-kx/2)}{1-\exp(-kx)}\right] \qquad (4.12)$$

where $x = b_1 d$, $y = b_1 r_c$ and $r_c$ is the parameter of the pseudopotential. The values of $r_c$ are calculated starting from the condition that the pressure of the electron gas equals zero [31]. $r_c = r_c(n_+, \varepsilon)$. The data on single crystals from [32] are used to calculate $\gamma_M$ and $\gamma_{ps}$.

It is possible to express the work function for the (010) plane as

$$\varphi = -E_+ - \varphi^{(0)}(0) + \varphi_{ps} \qquad (4.13)$$

where $E_+$ is the energy of the uniform electron liquid per electron, $\varphi^{(0)}(0)$ is the value of the electric potential on the surface:

$$\varphi_1^{(0)}(0) = -\frac{\pi^3 n_+}{3b_1^2}; \qquad \varphi_2^{(0)}(0) = -\frac{2\pi n_+}{b_2^2};$$

$$\varphi_{ps1} = \frac{b_1}{n_+} \sum_{k=1}^{\infty} \frac{(-1)^{k+1}}{k^2} I_6; \qquad \varphi_{ps2} = \frac{b_2}{n_+} \gamma_{ps2}$$

The data obtained are presented in Table 4.2.

**Table 4.2** The effect of the elastic strain $\varepsilon$ on the surface energy $\gamma$ and on the work function $\varphi$ for the model single crystal of aluminum.

| $\varepsilon$ | 0 | 0.01 | 0.02 | 0.03 |
|---|---|---|---|---|
| $b^{-1}$ (pm) | 86.31 | 85.78 | 85.25 | 84.72 |
| $\gamma$ (J m$^{-2}$) | 0.404 | 0.407 | 0.410 | 0.413 |
| $\varphi$ (eV) | 3.346 | 3.354 | 3.363 | 3.371 |
| $\varphi_{exper}$ (eV) | 3.500 | 3.517 | 3.535 | 3.540 |

The results are almost independent of the choice of the test function $n(z)$, (4.6) or (4.7). The work function increases in all instances if the elastic strain rises. The computed value of the work function increases by value of the order of 25 meV if the deformation of aluminum is 3%. In our experiments we observed an increase of 40 meV. Authors of [24] reported an increment of 30 meV. It is clear that the results of calculations are in an agreement with the experimentally observed values for aluminum. However, the obtained values of the surface energy seem to be too low.

The electron layer near the metallic surface is a 0.086 nm thick, which is $\approx 0.3$ of the interatomic distance.

The basic causes for an increase in the work function under elastic strain are a gradual decrease in the electron concentration $n_+$ and a decrease in the interplanar spacing $d$. This leads to a decrease in the $b^{-1}$ value of $\approx 1.8\%$, see Table 4.2. It appears as though the external electron tail moves in the elastic strained crystal, closer up to the surface.

## 4.2
### Taking into Account the Relaxation and Discontinuity of the Ionic Charge

LEED data have been published that indicate a displacement of the topmost layer relative to the bulk. A changing in the interplanar distance $d$ in the surface plane has been reported by several researchers [34–36]. That is to say the surface crystal lattice relaxes (see Figure 1.1).

We now try to take into account the discontinuity in the positive charge and the relaxation of the crystal planes at the surface.

In the next model, the charge of every ion is considered to be smeared over the planes, which are parallel to surface [33]. The charges are concentrated in planes of thickness $d$, where $z = -\frac{1}{2}d, -\frac{3}{2}d, -\frac{5}{2}d, \ldots$ (Figure 4.2). The size of the ion is determined by a parameter $r_c$ which is the Ashkroft pseudopotential, where $r_c < \frac{d}{2}$. The ion–electron interaction is neglected inside the ion area.

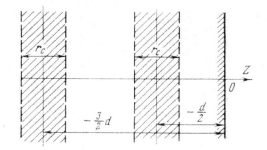

**Fig. 4.2** The one-dimensional model of the discrete ion lattice for a metal.

In this case the surface energy consists of three components,

$$\gamma = \gamma_0 + \gamma_{cl} + \gamma_{ps} \tag{4.14}$$

where $\gamma_0$ is the energy of the electron system, $\gamma_{cl}$ is the energy of interaction of the ions, $\gamma_{ps}$ is related to the energy of the electron–ion interaction.

For calculations of the work function increment in elastic strained metals the variational problem has been solved [90]. Two variational parameters were determined starting from a minimum of the surface energy. The parameter $\lambda$ determines a coordinate of the first relaxed plane: $z_1 = -d(0.5 - \lambda)$. Another parameter $b$ is related to the electron distribution, (4.6).

The results of calculations are shown in Figure 4.3. The elastic strain results in an increase in the work function. The experimental data for aluminum (curve D) are in close agreement with the results of the discrete ion model. However, the model fails for nickel specimens.

The value of the work function is extremely dependent upon the crystal plane. The model under consideration gives the anisotropy for the face-centered cube lattice of aluminum as

$$\varphi_{[100]} < \varphi_{[111]} < \varphi_{[110]}$$

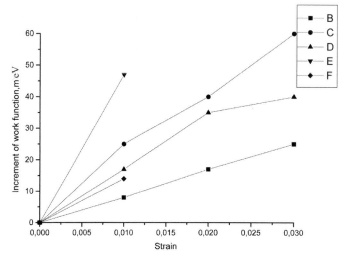

**Fig. 4.3** Dependence of the work function increment on the elastic strain: B, aluminum, the jellium model; C, aluminum, the model of the ionic discrete lattice with topmost relaxation; D, experimental data for polycrystalline aluminum; E, nickel, the model of the ionic discrete lattice with topmost relaxation; F, experimental data for Ni.

## 4.3
## Model for Neutral Orbital Electronegativity

Our aim is to develop a method for the work function calculation of the strained metal surface, which is based on the semi-empirical theory for neutral orbital electronegativity.

### 4.3.1
### Concept of the Model

A surface atom is assumed to conserve its individuality in the model of neutral orbital electronegativity [37]. Wave functions describing the state of electrons on the surface are approximated for a superficial atom by the wave functions of the isolated atoms. Thus, on a surface there will be some atomic orbitals or their combinations. The electronegativity is equal to the chemical potential of the orbital but with the opposite sign. Each removed electron was situated at the orbital of the surface atom. The work function varies considerably with strain as the surface potential changes. These changes in the potential depend on the surface microscopic geometry and also on the coordination number of atoms. The experimental determination of the surface atom coordination has become possible due to the achievements in scanning tunnel microscopy [13, 38].

Numbers of disrupted interatomic bonds with nearest and with most distant neighbors $i$, $j$ respectively, are used for characteristics of the surface imperfection. The dependence of the work function $\varphi_{ij}$ on $i$ and $j$ is expressed as [39]

$$\varphi_{ij} = x_{ij} = 0.98\frac{(V_n - i)n_a + (V_{nn} - j)n_b + 1}{r_a} + 1.57 \text{ eV}, \tag{4.15}$$

where $x_{ij}$ is the orbital electronegativity of an external atom on the surface, $V_n$ and $V_{nn}$ are numbers of the nearest and the next-nearest neighbors of atoms in the bulk, respectively, $V_n - i$ and $V_{nn} - j$ are numbers of bonds of the external atom with the nearest and the next-nearest neighbors, $n_a$ and $n_b$ are numbers of electrons taking part in bond of the surface atom with the nearest and distant neighbors referred to one atom, $r_a$ is the Pauling atomic radius of the element[1]. The coefficients 0.98 and 1.57 are fitting parameters.

The semi-empirical formula (4.15) is substantiated in [40]. This formula describes the experimental data well for the work function of crystal planes of metals. In Figure 4.4 one can see the ability of the model of electronegativity to represent the work function of metals. The data we have calculated are in close agreement with experimental results taken from [29].

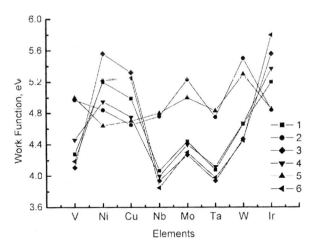

**Fig. 4.4** Values of the work function for some metals. 1, 2, 3, calculated by the method of neutral orbital electronegativity: 1, crystal plane (100), 2, (110), 3, (111); 4, 5, 6, experimental data from [29]: 4, crystal plane (100), 5, (110), 6, (111).

The average value of $\varphi$ was calculated as follows. The arrangement of atoms on the free surface of a metal is computed at any moment of strain. The defor-

---

[1] $r_a$ in units of 0.1 nm.

mation is determined as the function of the free surface increment. The work function is calculated at the change in the surface atomic order. The distances between the nearest and the next-nearest neighbors, $R_1$, and $R_2$, respectively, are changed as the result of the plastic strain. These values affect the number of electron bonds of the surface atoms with the nearest and the most distant neighbors, $n_a$, and $n_b$, respectively. The work function is calculated for the elastic and the plastic strain $\varepsilon$ for each value of $i$ and $j$. Calculations are performed for different crystal planes of a single crystal of copper and aluminum.

For the [100] direction in the cubic face-centered crystal lattice

$$R_1(\varepsilon) = a\frac{(2+\varepsilon^2+2\varepsilon-2\nu\varepsilon+\nu^2\varepsilon^2)^{\frac{1}{2}}}{3} + a\frac{1-\nu\varepsilon}{3\sqrt{2}} \tag{4.16}$$

$$R_2(\varepsilon) = \frac{2a(1-\nu\varepsilon)+a(1+\varepsilon)}{3} \tag{4.17}$$

where $a$ is the crystal lattice parameter, $\nu$ is the Poisson coefficient and $\varepsilon$ is the strain.

The number of electrons that binds the surface atom with its nearest neighbors is given by

$$n_a(\varepsilon) = \frac{\nu}{V_n + V_{nn}\exp\left[\frac{R_1(\varepsilon)-R_2(\varepsilon)}{0.26}\right]} \tag{4.18}$$

The number of electrons that binds the surface atom with its next-nearest neighbors can be expressed as

$$n_b(\varepsilon) = n_a(\varepsilon)\exp\left[\frac{R_1(\varepsilon)-R_2(\varepsilon)}{0.26}\right] \tag{4.19}$$

The double electric layer on the metal surface is the layer of dipoles (see Section 1.4.2, Figure 1.14). The elastic strain leads to a redistribution of the electric charge. The electrostatic surface barrier and the work function change accordingly. Taking into account the correction due to the charge redistribution (4.15) can be rewritten as

$$\varphi_{ij} = \left[0.98\frac{(V_n-i)n_a+(V_{nn}-j)n_b+1}{r_a}+1.57\right]\cdot\left[1+\frac{\delta(\varepsilon)-\delta(0)}{\delta(0)}\right] \tag{4.20}$$

where

$$\delta(\varepsilon) = \frac{1}{\pi}\cdot\left[\arctan\frac{\rho+R_1(\varepsilon)/2}{\lambda} - \arctan\frac{\rho-R_1(\varepsilon)/2}{\lambda}\right] \tag{4.21}$$

where $\lambda$ is the thickness of the double electric layer and $\rho$ is an effective length [39].

## 4.3 Model for Neutral Orbital Electronegativity

Let us assume the reasonable quantities of $\lambda = 0.05$ nm, $\rho = 0.05$ nm. Varying the values of $i, j$ we can calculate the increment of the work function from (4.20) for different crystal planes.

The data obtained for aluminum are illustrated in Figure 4.5. The theoretical dependence is in satisfactory agreement with the experimentally obtained values of the work function for elastic strained aluminum (Table 4.2, Figure 4.3, curve D).

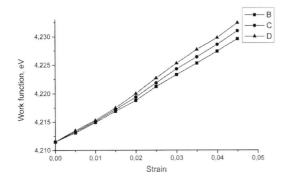

**Fig. 4.5** The work function of the elastic strained aluminum calculated according to the model of the neutral orbital electronegativity: B, crystal plane (110); C, crystal plane (100); D, plane (111). The curves were superposed in order to be coincided at $\varepsilon = 0$.

### 4.3.2
### Effect of Nanodefects Formed on the Surface

The appearance of dislocation steps and more complicated defects on the surface changes the electrostatic superficial barrier and has an influence on the work function. The influence is dependent on the density of the surface imperfections.

The deformation processes have been shown [13, 38] to be related to the initiation and evolution of nanometric defects. These defects have a prism shape. The sides of the prisms are formed due to emerge the dislocations on the surface along planes where slipping is easy. A prism of this kind is shown in Figure 4.6.

We have calculated the work function for plastic strained copper in accordance with (4.16)–(4.21).

If $k$ nanodefects lined up across a specimen they would cause an elongation $dl$. If the number of lines is $n$ then the elongation of the specimen is $n \cdot dl$. We can find a relationship between the number of nanodefects and the specimen strain. Steps from 5 to 50 nm wide are formed when the dislocations appear and 59 dislocations were found to generate a step of 15 nm high [41]. It was

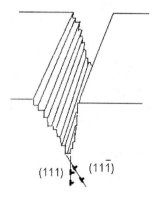

**Fig. 4.6** A surface defect of prism shape formed by slipping dislocations. Dislocations slide along planes of type {111}.

observed that nanodefects on the surface of the loaded copper form four ensembles. They differ considerably in dimensions. The dimension of the next defect is three times as much as that of the previous one [42].

The length of the sides of the nanodefects of first rank in copper is equal to approximately 80 nm. They are formed by the emission of $\approx$ 300 dislocations [13]. The density of defects of first rank increases until the entropy of admixture of defects and atoms of the crystal lattice achieves a maximal value. Part of the defects dissolves later, an another part forms nanodefects of second rank. The density of the second-rank defects increases up to $\approx$ 5 %. Later they dissolve and form partially defects of the third rank. The time-dependent oscillation of nanodefects was included into the program of calculations.

The decrease in the work function under elastic strain was concluded to be determined by the initiation of surface defects of first rank. The calculated work function of these defects in copper is found to be 3.999 eV. The influence of defects of rank 2, 3, 4 on the work function value, appears to compensate for its decrease to some extent. The calculated values of the work function are equal to 4.028, 4.038, 4.049 eV for defects of rank 2, 3, and 4, respectively.

The results of the calculations are illustrated in Figure 4.7. The decrease in the work function is represented by a theoretical curve B and experimental curves C and D. The fit between the results of modeling and the experimental data seems to be satisfactory.

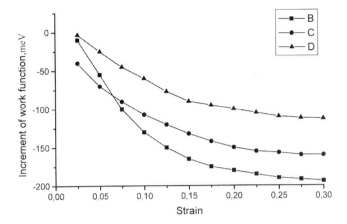

**Fig. 4.7** Dependence of the increment of the work function on the elastic strain for copper: B, the curve computed in accordance with the model of broken bonds; C, the results were calculated from the experimental data of [43]; D, experimental data from [24]

## 4.4 Summary

We have described models for the work function of the strained metallic surface. We used the jellium model to investigate the effect of the strain on the surface energy and the electron work function. In this model the positive charge of ions is distributed uniformly over the volume. Free electrons are moving in the uniform positive background created by metallic ions.

The elastic strain of a single crystal is assumed to be in a preferred direction. This leads to a change in the atomic density of the parallel planes, the distance between them, and a variation in the electron concentration.

We consider the elastic strain along the [100] direction of a crystal with the cubic face centered crystal lattice. Defining the distribution of the electron density along the [010] axis by one-parametric probe functions $n(z)$ we obtain the parameters from the condition of minimum surface energy.

The work function increases in all cases if the elastic strain increases. The calculated value of the increase in the work function is of the order of 25 meV for 3% deformation of aluminum. The experimental result is 40 meV. The calculations are in agreement with the experimentally observed values for aluminum.

The electron layer near the metallic surface is equal $\approx 0.3$ of the interatomic distance. The basic cause for an increase in the work function for an elastic strained crystal is the gradual decrease in the electronic concentration, and in the interplanar spacing. The external electron tail is located in the elastic strained metal closer up to the surface than in unstrained metals.

Further, we have taken into account the discontinuity of the positive charge and the relaxation of the crystal planes at the surface. In our next model the charge of every ion is considered to be smeared over planes of thickness $d$, which are parallel to the surface. The size of the ion is determined by a parameter $r_c$, that is, the Ashkroft pseudopotential, $r_c < \frac{d}{2}$. The ion–electron interaction is negligible inside the ion area. In this case the surface energy consists of three items: the energy of the electron system, the energy of interaction of the ions, and the energy of the electron–ion interaction. For calculations of the increment in the work function in elastic strained metals the variational problem has been solved. Two variational parameters were determined starting from a minimum of the surface energy. The elastic strain results in an increase in the work function. The results for the discrete ion model are in good agreement with experimental data for aluminum. The value of the work function is extremely dependent upon the crystal plane.

$$\varphi_{[100]} < \varphi_{[111]} < \varphi_{[110]}$$

The semi-empirical theory of neutral orbital electronegativity was also used to estimate the work function of strained surfaces. A surface atom was assumed to conserve its individuality in this model. Wave functions describing the state of electrons on the surface are approximated by wave functions of the isolated atoms. Thus, on a surface there are some atomic orbitals or their combinations. The electronegativity is equal to the chemical potential of the orbital but with the opposite sign. The work function varied considerably with strain as the surface potential changed abruptly. These changes in the potential depend upon the surface microscopic geometry and upon the coordination number of atoms.

The deformation processes were shown to be determined by the initiation and evolution of nanodefects. These are nanometric defects having the shape of prisms. The sides of the prisms are formed by the dislocations appearing on the surface over planes where sliding is easier.

Nanodefects on the surface of the loaded copper form four ensembles. The dimension of a subsequent defect is three times greater than that of the previous one. The length of the sides of nanodefects of first rank in copper is 80 nm. These are formed at the emission of $\approx 300$ dislocations. The density of the defects of first rank increases with strain. Then part of defects dissolves, and another part forms nanodefects of the second rank. The density of second rank defects increases up to $\approx 5\,\%$. Later they dissolve and form defects of third rank.

Calculations of the work function for copper surface with nanodefects have been performed based on the theory of orbital electronegativity. Time-dependent oscillation of nanodefects was included into the program of calculations. It was observed that the work function decrease under plastic strain is mainly determined by the formation of surface defects of first rank. The calculated work function is found to be equal to 3.999 eV. The calculated values of the work function are 4.028, 4.038, 4.049 eV for defects of rank 2, 3, and 4, respectively.

The fit between the results of modeling and the experimental data seems to be satisfactory.

# 5
# Contact Interaction of Metallic Surfaces

Contact interactions of metal surfaces are inherent in processes of friction, mechanical treatment and strengthening. Physical and mechanical characteristics of surface layers are important for the wear resistance and durability of machine components. It is the object of this chapter to consider the data on interaction of metal surfaces during their relative movement. The technique of local indentation is effective method for measurement material characteristics. We also consider how the surface roughness and wear influence the energetic relief of the strained surface.

## 5.1
## Mechanical Indentation of the Surface Layers

Elastic and plastic strain both occur during the loading of a contact pair. Elastic restoration takes place during unloading of the contact area. Therefore, processes occurring during the unloading are more easily explained since they include only the elastic component of the deformation.

The problem for elastic contact of two solids is called the Hertz problem. The elastic approach of a spherical body (indenter) to a plane surface is expressed as [44]

$$h_{el} = P^{\frac{2}{3}} D^{\frac{2}{3}} \left( \frac{1}{r} + \frac{1}{r_1} \right)^{\frac{1}{3}} \tag{5.1}$$

where $h_{el}$ is the elastic approach, $P$ is the load, $D$ depends on elastic moduli, $r$ is the radius of the indenter and $r_1$ is the radius of the curvature of the plane, $1/r_1 \to 0$.

The value of $D$ is given by

$$D = \frac{3}{4} \left( \frac{1 - \nu_s^2}{E_s} - \frac{1 - \nu_i^2}{E_i} \right) \tag{5.2}$$

where $\nu_s$, $E_s$, and $\nu_i$, $E_i$ are the Poisson coefficients and the Young moduli of the plane specimen and of the indenter, respectively.

*Strained Metallic Surfaces.* Valim Levitin and Stephan Loskutov
Copyright © 2009 WILEY-VCH Verlag GmbH & Co. KGaA, Weinheim
ISBN: 978-3-527-32344-9

The radius of the imprint $\rho$ is given by

$$\rho = P^{\frac{1}{3}} D^{\frac{1}{3}} \left( \frac{1}{r} + \frac{1}{r_1} \right)^{-\frac{1}{3}} \tag{5.3}$$

Combining (5.1) and (5.3) we obtain

$$h_{el} = \frac{PD}{\rho} \tag{5.4}$$

The mean contact stress $\bar{\sigma}$ is expressed as

$$\bar{\sigma} = \frac{P}{\pi \rho^2} \tag{5.5}$$

Thus, the elastic approach of the bodies during unloading can be described as

$$h_{el} = D(\pi \bar{\sigma} P)^{\frac{1}{2}} \tag{5.6}$$

It is possible to express the elastic and plastic strain that occur during the loading as [45]

$$h_{el-pl} = \frac{2P}{\pi C^2 r \bar{\sigma}} \tag{5.7}$$

where the coefficient $C = 1.98$.

One should distinguish the contour area of contact $A_c$ and the actual area of contact $A_a$ where $A_a < A_c$. Let us denote the ratio of these areas as

$$\alpha = \frac{A_a}{A_c} \tag{5.8}$$

Assuming that the approach of bodies, $h$, is the same provided that $A_a$ is the same, for a rough surface and for a smooth surface we have

$$A_a = \alpha 2\pi r h_{el} = \alpha 2\pi r h_{el-pl} \tag{5.9}$$

We have measured the electrical resistance, $R$, of the contact pairs the spherical indentor – the surface of the specimen, see Section 2.4 and Figure 2.16.

The contact electrical resistance of a pair is inversely proportional to the actual contact area. For two actual areas, $A_c$, and $A_0$, we note that

$$\frac{R_0}{R} = \frac{A_a}{A_0} \tag{5.10}$$

Let us denote

$$\beta = A_0 R_0 \tag{5.11}$$

Equations (5.6)–(5.11) can be combined for both the elastic and plastic loading to obtain the expression

$$\frac{\beta}{R} = \frac{4\alpha P}{C^2 \bar{\sigma}} \tag{5.12}$$

For elastic unloading, the expression is

$$\frac{\beta}{R} = 2\pi\alpha Dr(\pi\bar{\sigma}P)^{\frac{1}{2}} \tag{5.13}$$

It follows from (5.12) and (5.13) that a linear dependence $1/R \sim P$ can be expected for elastic–plastic indentation (that is, during the loading). One can expect the proportionality $1/R \sim \sqrt{P}$ in the case of the elastic restoration of the imprint (during unloading).

The results of measurement of the electrical conductivity of the pair the steel ball – the plane surface during the loading are presented in Figure 5.1[1]. The linear dependence $1/R - P$ is observed. At the first loading the curve consists of two linear segments. The straight-line segment $I$ corresponds to elastic strain, whereas the segment $II$ corresponds to elastic and plastic deformation. At segment $II$ the central part of the contact area is deformed plastically. The deflection of the imprint increases and the contact stress decreases. The change in the inclination of the curve 1 is due to a smoothing of the peaks on the surface profile.

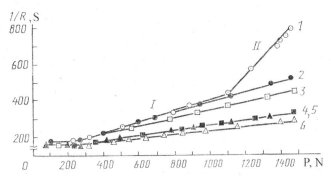

**Fig. 5.1** The electric conductivity of the conjugated pair versus the applied load. A ball-bearing of diameter 23.65 mm contacts with the plane specimen of the titanium-based VT3-1 alloy. Curve 1 corresponds to the first loading, curves 2–6 correspond to subsequent loadings with intermediate unloadings. The factor of the initial roughness $R_a = 1.01$ μm; coefficient $\beta = 25.2 \times 10^{-10}$ Ω · m².

The mean contact stress increases during subsequent loadings, Figure 5.1, curves 2–6. The value of $\bar{\sigma}/\alpha$ increases from 1.32 (curve 1) to 3.61 (curve 6) GPa.

[1] The reciprocal Ohms, Siemens, are plotted on the ordinate axis.

Figure 5.2 illustrates subsequent unloadings of the contact pair. The dependence $1/R \sim \sqrt{P}$ is indeed a linear one. Subsurface layers are elastically restored during the first unloading (the segment II, Figure 5.2). Also the restoration of the rough surface begins. The value of the mean contact stress decreases in accordance with (5.13). The smaller the roughness of the surface $R_a$ the greater the value of the mean contact stress, $\bar{\sigma}$.

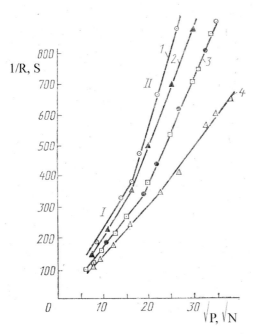

**Fig. 5.2** Dependence of electric conductivity on $\sqrt{P}$ for the conjugated pairs during unloading. Specimens 1, 2, 3, were treated by different emery papers; specimen 4 was treated previously by ball-bearings in an ultrasonic field. The roughness of the surface of specimens, $R_a$, is equal to (μm): 1, 0.10; 2, 0.39; 3, 1.01; 4, 0.74.

It is possible to express the actual contact area $A_a$ by means of an empirical equation as

$$A_a \simeq P^m \tag{5.14}$$

where the constant $m$ is dependent on the material. The value of $m$ in (5.14) was found to vary from 1/2 to 1.

In Figure 5.3 the dependence of the electrical conductivity on the square-root of the load is shown. Measurements were conducted with and without application of vibration. The rate of the conductivity increment $\frac{d(1/R)}{d(\sqrt{P})}$ increases with application of the vibration (Figure 5.3, curve 1). The kink in this curve is caused by an exhaustion of the plastic deformation. Curve 4 has

## 5.1 Mechanical Indentation of the Surface Layers

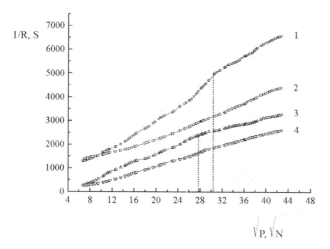

**Fig. 5.3** Electrical conductivity of contact pairs as a function of $\sqrt{P}$: 1, the initial surface and vibration during the test; 2, the preliminary compression of surfaces to the yield limit and the applied vibration; 3, the indentation of the initial surface; 4, the preliminary compression of the surface to the yield limit.

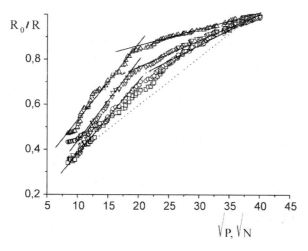

**Fig. 5.4** The dependence of $\frac{1/R}{1/R_0}$ on $\sqrt{P}$: $\triangle$ and $\triangledown$, unloading without vibration; $\square$ and o, unloading with vibration; the dotted line shows the same dependence for a smooth surface, which is obtained from the solution of the Hertz problem.

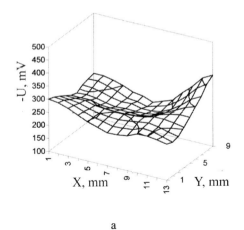

a

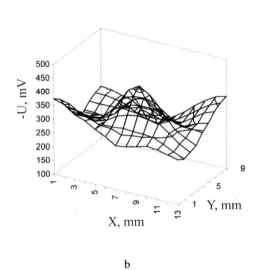

b

**Fig. 5.5** The distribution of the contact potential difference over the surface of the titanium-based VT 3-1 alloy: a, before indentation; b, after indentation by a polished ball-bearing of 57 mm diameter.

the smallest angle of inclination because the specimen was strained before the test.

Figure 5.4 illustrates the processes during unloading. The ratio of the initial conductivity to the current conductivity, $\frac{1/R}{1/R_0}$, is plotted along the ordinate axis. One can see two line segments. For the first one a relatively slow decrease in the value of $R_0/R$ is inherent. The elastic restore of the contour area of the imprint seems to be the only cause of this segment. At the second line-segment the conductivity drops quickly. This is related to the restoring of peaks of the surface roughness.

## 5.2
## Influence of Indentation and Surface Roughness on the Work Function

The technique of the work function allows one to observe the plastic flow of metal during indentation. Figure 5.5 illustrates the space distribution of the contact potential difference over the surface of the imprint. A ball-bearing of diameter 57 mm is used for indentation. The step of the work function scanning is equal to 200 µm.

The central part of the imprint in Figure 5.5b corresponds to a region of severe plastic deformation. The area is spherically symmetric.

One can see sections of the contact potential distribution in Figure 5.6. The increment in the work function at the center of the imprint is equal to $-400$–$-500$ meV. The strained area of the imprint becomes wider if the load increases.

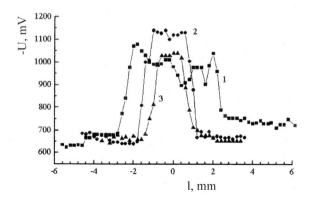

**Fig. 5.6** The contact potential difference versus distance across the imprint of the ball-bearing on the plane specimen of the titanium-based alloy. 1, the load equals to 1500 N; 2, 700; 3, 300 N.

The effect of the surface roughness on the work function has been investigated. Specimens of aluminum have been annealed at 523 K in a vacuum $1.33 \times 10^{-1}$ Pa for 4 hours. The mean value of the work function is equal to 3.40–3.50 eV. The surface of each specimen is then divided into five parts. Each part is burnished using the different emery paper. In order to exclude the influence of other factors, we conduct all treatment within the surface of the same specimen. The parts with different roughness are located side by side. The mean hight of peaks $R_a$ varies from 0.10 to 1.45 µm. Values of $R_a$ are plotted on the abscissa axis in Figure 5.7.

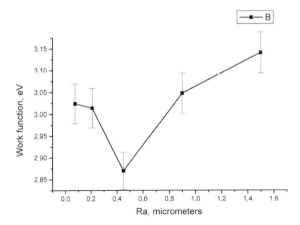

**Fig. 5.7** The electron work function versus the mean height of peaks on the surface of aluminum at once after grinding or polishing.

The dependence $\varphi - R_a$ has a minimum at $R_a = 0.45$ μm. A decrease and also an increase in the surface roughness leads to a growth in the work function.

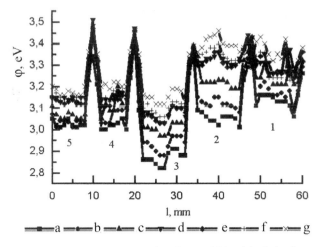

**Fig. 5.8** Effect of surface grinding (1 - 4) or polishing (5) of aluminum on the electronic work function. The $R_a$ values are (μm): 1, 1.50; 2, 0.90; 3, 0.45; 4, 0.21; 5, 0.075. Boundaries between specimen parts have initial values of $\varphi = 3.40 - 3.50$ eV. The curves show: a, immediately after grinding or polishing; b, after 1.5 hours; c, after 20 hours; d, after 70 hours; e, after 73 hours; f, after 120 hours; g, after 240 hours.

One can see more detailed results in Figure 5.8. For the polished area (number 5) the work function drops to 3.0 eV; a minimum of 2.85 eV occurs at region 3 where $R_a = 0.45$ μm.

Relaxation processes occur differently depending on the surface roughness. The work function of areas having a rough surface (1, 2) returns to initial values 3.40–3.50 eV. The regions 3–5 do not reach these values, remaining at the level 3.2 eV.

## 5.3
## Effect of Friction and Wear on Energetic Relief

Figure 5.9 shows the work function of the surface friction along ring tracks. The tests are carried out at relative velocities 0.006 and 0.011 m s$^{-1}$ under a load of 1.5 N. A spherical indentor of 3 mm in diameter is used. The work function decreases along friction tracks for all the materials under study. Distances between minimums are in accordance with the size of ring tracks.

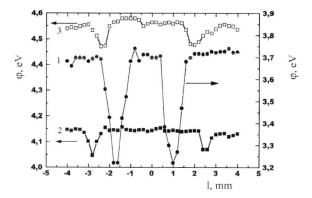

**Fig. 5.9** The electron work function measured along friction tracks: 1, aluminum; 2, a hard metal of the system Fe-C-Cr-B; 3, the titanium-based VT 3-1 alloy.

One can see the work function kinetics during the friction test in Figure 5.10. The work function drops sharply from $-100$ to $-280$ meV during the first hour. Then a decrease up to $-300$ meV takes 19 hours. It is obvious that the formation of a new surface structure occurs in the initial time interval of the friction process. One labels this process as the running-in period.

It is important that the work function is sensitive to processes of friction and wear. The fact is that the wear in a real tribological system is less than it is in experimental devices. This leads to a difference of about three to five orders of magnitude between reality and a the model [46]. Friction and wear result in induced changes in the surface. After friction the fragmented structure is formed in the surface layers of the metal (Figure 5.12).

A wave-like structure very often develops on surfaces in tribological interactions. Figure 5.11 shows the microstructure of contact surfaces after wear

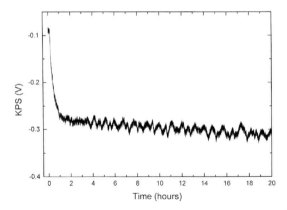

**Fig. 5.10** The electronic work function (Kelvin probe signal, KPS) versus time. A contact pair of copper – bronze, load 4,32 N. Reprinted from [47] with permission from Elsevier.

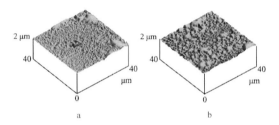

**Fig. 5.11** Images of two worn cast iron surfaces: a, the contact pressure is 30 MPa; b, the contact pressure is 90 MPa. Reprinted from [46] with permission from Elsevier.

tests. A long-wave structure appears under a high load. The short-wave structure is intrinsic to a relatively low load. The rough relief causes electron redistribution near the surface and leads to a decrease in the work function.

The evolution of the fragmented structure in surface layers under friction is shown schematically in Figure 5.13.

Tribological properties of the surface are shown to be essentially dependent on the size of structural constituents. A surface layer with grain sizes on the nanometric scale ensures a good wear strength. A typical electron diffraction pattern of the nanostructured surface layer is presented in Figure 5.14.

Figure 5.15 illustrates the dependence of the friction coefficient on nanostructuring of surface. The friction coefficient is sufficiently decreased by severe plastic deformation of the surface.

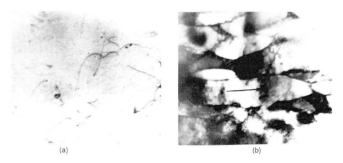

**Fig. 5.12** Structure of surface layers of a low carbon steel: a, before friction; b, after friction at a sliding distance of $5 \times 10^3$ m under a pressure of 0.3 MPa. The pattern of a transmission electron microscope; the magnification was not reported. Reprinted from [48] with permission from Springer Science.

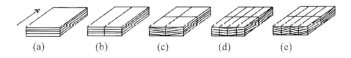

**Fig. 5.13** Scheme of formation of fragmented structure during the running-in period. Reprinted from [48] with permission from Springer Science.

The authors of [50] have attempted to find a logical connection between the electron work function and processes of adhesion and friction under a light load. Their reasoning is as follows.

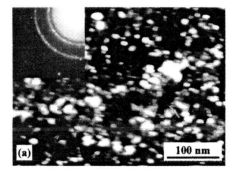

**Fig. 5.14** Dark-field electron micrograph showing reflexes from crystallites of 10 - 20 nm dimension. Inset is the electron-diffraction pattern that indicates solid reflexes. Reprinted from [49] with permission from Elsevier.

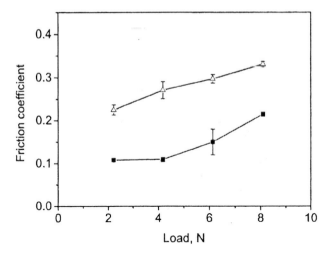

**Fig. 5.15** The friction coefficient versus the applied load for surfaces of low-carbon steel: △, non-treated metal; ■, surface after mechanical attrition treatment. Reprinted from Ref. [49] with permission from Elsevier.

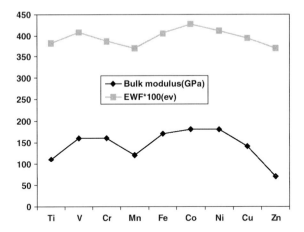

**Fig. 5.16** The dependence of the electronic work function (EWF) on the bulk modulus (after [50]).

The electron work function reflects the surface electronic behavior. This parameter is known to be a fundamental property of materials. Also, it is affected by a wide range of surface phenomena. On the other hand, mechanical properties are also related to electron behavior. The bulk modulus depends on the electron subsystem and the nature of the interatomic bond in the solid. The cohesive energy, elastic constant, and hardness are closely connected with the bulk modulus. Hence, one can assume the existence of a correlation between the electron work function and mechanical characteristics as well as between the work function and tribological parameters.

In Figure 5.16 this correlation is shown for some 3d-transition metals. The measured work function reflects the properties of a natural surface rather than that of a pure metallic surface. In spite of this fact one can detect a trend. The higher the bulk modulus of a metal the larger the work function.

According to the author's model, the work done against the friction force is consumed partially in breaking the adhesive bonds and in the elastic strain in surface layers of the contacting metals. The load is assumed to be relatively low. It is obvious that the authors of [50] are neglecting plastic strain in this model.

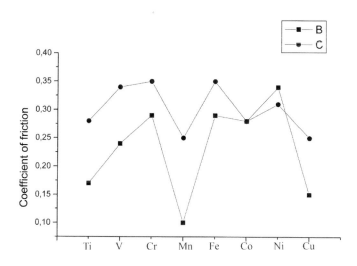

**Fig. 5.17** Comparison of data for 3d-metals: B, measured coefficients of friction, contact pairs are a zinc 4 mm semi-sphere and a metallic plate under a load of 1 mN; C, calculated coefficients of friction. Data from [50].

Thus, the performed work can be expressed as

$$F \cdot S = (E_s \cdot A + U_E \cdot A) \frac{S}{d} \tag{5.15}$$

where $F$ is the friction force, $S$ is the total distance of sliding, $E_s$ is the adhesion energy per unit area, $U_E$ is the energy of the elastic strain for unit contact area, $d$ is the diameter of the contact area, and the relation $S/d$ is the number of sliding steps. The friction coefficient $\mu$ by definition is

$$\mu = \frac{F}{L} \tag{5.16}$$

where $L$ is the normal load.

The authors measured the friction coefficient for contact pairs consisting of a zinc semi-sphere of 4 mm radius and plates mode of different metals.

Three key parameters are considered necessary in order to calculate the coefficient of friction. These are: the contact area; the adhesion force; and the elastic energy. The strain energy was estimated to be negligible for elastic contact under light loads. Under a load of 1 mN the strain energy of a semi-sphere of 4 mm in contact with a plate is estimated to be of the order of $10^{-3}$ J m$^{-2}$, which is sufficiently small in comparison with the adhesion energy, (1 J m$^{-2}$). The contact area $A$ is a function of the load, the radius of the semi-sphere, and the Young modulus of the pair. The adhesion energy $E_s$ for a pair of metals is found to be the sum of surface energies of both metals minus their interfacial energy. We refer the reader for details of calculations to [50] and to references therein.

Figure 5.17 presents a comparison of calculated and measured friction coefficients. As one would expect, the model cannot predict the value of the friction coefficient for a specific metal. However, the trend of the $\mu$ variation for different metals seems to be correct.

## 5.4
## Summary

Elastic and plastic strain occur during the process of loading the contact pair. During unloading the elastic restoration of the imprint takes place. Processes occuring for a decrease in the load include only the elastic component of the deformation.

One should distinguish between the contour area and the actual area of a metallic pair contact. The electrical conductivity of a contact pair is inversely proportional to the actual contact area.

A linear dependence $1/R \sim P$ was observed at elastic–plastic indentation (that is, when the load $P$ increases). During the first loading the straight-line segment $I$ corresponds to an elastic strain, whereas the segment $II$ is due to elastic and plastic deformation. During subsequent loadings of the contact pair the first curve $1/R - P$ consists of two linear parts. At segment $II$ the central part of the contact area was deformed plastically. The deflection of the imprint increases, the contact stress decreases. The change in the slope of the straight line dependence is due to smoothing of the peaks on the surface profile. The mean contact stress increases during subsequent loadings.

One observed the proportionality $1/R \sim \sqrt{P}$ in the case of unloading and elastic restoration of the imprint. During unloading one could see two linear sections at curves $\frac{1/R}{1/R_0}$. For the first one a, relatively slow, decrease in the value $R_0/R$ is inherent. Elastic restoration of the contour area of the imprint seemed to be the only the of this segment. At the second segment the conductivity drops quickly. This is related to the restoration of the peaks of surface roughness.

The technique of the electron work function allows one to observe the plastic flow of metal during indentation of a contact pair a ball – a plane specimen. The central part of the imprint corresponds to a zone of severe plastic deformation. The area is spherically symmetric. The increment in the work function at the center of the imprint was about $-400$ to $-500$ meV. The greater the load the wider the strained area of the imprint.

Friction and wear resulted in induced changes on the surface. During friction a fragmented structure and a wave-like surface were formed.

The work function procedure was found to be sensitive to processes of friction and wear. The work function decreased along friction tracks for all materials under study.

Tribology properties of the surface were shown to be essentially dependent on the size of the structural constituents. A surface layer with grain sizes on the nanometric scale ensure a good wear strength. The friction coefficient of the nanostructured surface was sufficiently decreased compared with the ordinary grain size.

Attempts have been made to find a connection between the electronic work function and friction under a light load. The results obtained could not predict the value of the friction coefficient for a specific metal. However, the trend of the variation in the friction coefficient for different metals seems to be correct.

# 6
# Prediction of Fatigue Location

A metal which is subjected to repetitive or alternating stresses fails at a much lower stress than that required to cause a fracture on the single application of a load. Failures occurring under conditions of oscillating loading are called fatigue failures. Fatigue fracture is generally observed after a considerable period of time.

The basic mechanism of fatigue fracture is the origination of a crack on the surface which slowly spreads. Fatigue becomes progressively more prevalent as technology develops a greater amount of equipment, such as turbines, aircrafts, automobiles, compressors, pumps, bridges etc., which are subjected to repeated loading and vibration. Material fatigue is known to be a very dangerous type of fracture. Today it is often stated that fatigue accounts for at least 90 percent of all service failures which are due to mechanical causes.

Fatigue failure is particularly insidious because it occurs without any obvious warning. Fatigue results in a brittle-appearing fracture, with no gross deformation at the fracture. On a macroscopic scale the fracture surface is usually normal to the direction of the principal tensile stress. Fatigue failure can usually be recognized from the appearance of the fracture surface, which shows a smooth region, due to the rubbing action as the crack propagates through the section. A second region of the fracture is rough because the load-bearing member fails in a ductile manner when the cross-section is no longer able to carry the load. Frequently, the progress of the fracture is indicated by a series of rings, or beach marks, progressing inward from the point of initiation of the initial crack.

In Chapter 3 we observed that plastic deformation results in a decrease in the work function measured at the surface of metals and alloys. On the other hand, plastic deformation events are known to precede fatigue fracture. Our aim in this chapter is to investigate the effect of fatigue tests on the work function in metals and alloys.

*Strained Metallic Surfaces*. Valim Levitin and Stephan Loskutov
Copyright © 2009 WILEY-VCH Verlag GmbH & Co. KGaA, Weinheim
ISBN: 978-3-527-32344-9

## 6.1
### Forecast Possibilities of the Work Function. Experimental Results

We measure the work function between bending cycles, that is, before, during and after fatigue tests. Measurements are taken along six longitudinal lines, which are parallel to the specimen axis. Experimental points are located 1 mm from each other. Thus, 40 measurement points were situated on the specimen gauge. Three measurements per point are performed and the average result is used. The ultraviolet radiation of specimens is applied in order to avoid the adsorption of gas molecules.

### 6.1.1
### Aluminum and Titanium-Based Alloys

Figure 2.18 1 and 6 show the shape of specimens of aluminium that are used for fatigue tests, Figure 2.18 2 illustrates specimens of titanium-based alloys.

The angular frequency of the tests, $\omega$, is 5881 rad s$^{-1}$ and 400 rad s$^{-1}$, respectively.

The work function is found to decrease at the very place where the crack would be formed, long before its creation. In Figure 6.1 the increment in the work function for the surface of aluminum is plotted against the number of cycles.

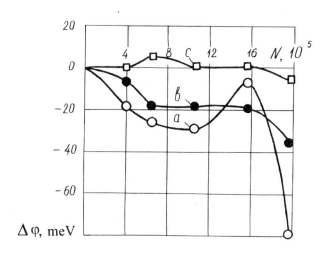

**Fig. 6.1** The increment in the electronic work function for an aluminum surface versus the number of cycles: a, area of a specimen directly above the future crack; b, 1 mm from that area; c, 3 mm from the same area.

This figure shows that an essential decrease in the work function occurs directly above the future crack (curve a). At the adjacent areas the decrease in the work function lags behind that of the former (curves b and c). The work function starts falling just after 5–10% of the total lifetime of the aluminum specimens.

It is important to emphasize that a decrease in the work function is also observed on the surface opposite to the one where a crack is being formed. Consequently, one can say that the observed drop in the work function is not caused by proper crack formation. This behavior of the work function indicates that it reflects structural changes on the specimen surface during the stress cycling.

One can see in Figure 6.2 that the greater the number of cycles the lower the drop in the work function.

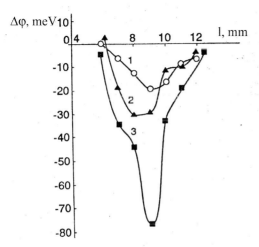

**Fig. 6.2** Distribution of the work function over the surface of aluminum during a fatigue test: 1, number of cycles $N = 1.0 \times 10^4$; 2, $N = 6.6 \times 10^5$; 3, $N = 2.0 \times 10^6$. A fatigue crack emerged at a distance of 9 mm from the fixing point of the specimen.

The distribution of $\varphi$ over the surface after testing the titanium-based VT 3-1 alloy is shown in Figure 6.3. The mean values of the work function plotted against the distance of specimens tested have a minimum value.

Near cracks (point 0) the decrease in $\varphi$ is $-41$ meV, far more than the measurement error ($\pm 1$ meV). The corresponding decrease for fatigued aluminum is equal to $-27$ meV.

Thus, the increment in the work function is found to be sensitive to structural changes during mechanical cycling of the specimens under investigation.

Figure 6.4 presents a correlation between the work function, the residual stresses, and the fatigue limit for the titanium-based alloy. We divided identical specimens into several groups. Specimens were strengthened by ball-bearing in the ultrasonic field. The amplitude of the vibrations and the treat-

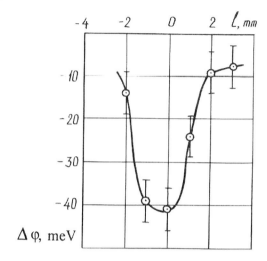

**Fig. 6.3** The distribution of the $\Delta\varphi$ value over the surface of the fatigued specimens. Every point is averaged over 16 specimens of the titanium-based VT 3-1 alloy. Zero is the point of crack formation.

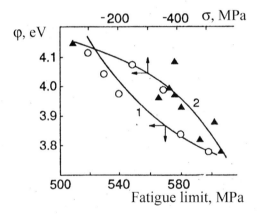

**Fig. 6.4** The correlation between values of the work function, the fatigue limit and the surface compressive residual stresses $\sigma$ for the titanium-based VT 3-1 alloy. Every point corresponds to a group of six specimens.

ment time varied from group to group. We averaged the data obtained over all specimens of the group.

The less the work function of treated specimens the higher is the fatigue limit. The work function of 4.1 eV corresponds to the fatigue limit of 520 MPa, 3.8 eV corresponds to that of 580 MPa. With the increase in compressive residual stresses the fatigue limit increases, while the work function decreases.

## 6.1.2
### Superalloys

Flat specimens of superalloys as shown in Figure 2.18, 3 were used; the test frequency is equal to 2338 rad s$^{-1}$. Considerable changes in the surface potential relief occur during the fatigue tests of superalloys.

The distribution of the work function for the EP479 superalloy after fatigue tests is presented in Figure 6.5. An area of 10 mm length from the left of the specimen is located at the clamp of the vibration source.

In an initial state, before fatigue tests the scattering of the work function quantities along specimens is observed (Figure 6.5, a). The scattering of points is about 30 meV. In Figure 6.5b one can see that the minimum of the work function appears after 50% of the specimen life at just $l = 20$ mm. At this point the work function drop is $-75$ meV. It is twice the scattering value before tests. One should compare the work function values in the load area (15–25 mm from the left) and the work function in the unloaded part (greater 30 mm). The largest level of applied cycling stresses is located at 17 mm from the left edge of the specimen. During subsequent fatigue tests, the minimum erodes,

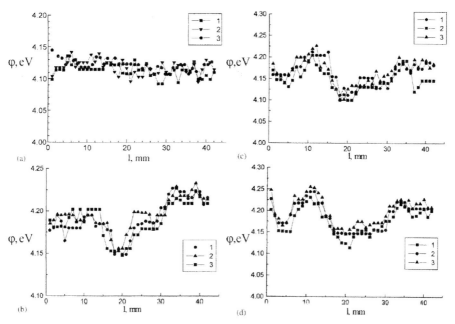

**Fig. 6.5** The distribution of the work function along fatigued specimens of the EP479 superalloy: a, initial values, the number of cycles $N = 0$, three kinds of points correspond to various series of measurements; b, curves from 1 to 3: $N$ is changed from 52 to $69 \times 10^6$ cycles; c, curves from 1 to 3: $N$ is changed from 72 to $84 \times 10^6$ cycles; d, curves from 1 to 3: $N$ is changed from 89 to $97 \times 10^6$ cycles.

Figure 6.5, c and d. The difference in the work function between unloaded and loaded areas of the specimen reaches $-100$ meV. The minimum at 5 mm in Figure 6.5d is caused by plastic deformation of the specimen surface with the clamp.

In Figure 6.6 the typical increment in the work function on the surface of an EP866 superalloy is plotted against the number of cycles. One can see that there is an evident down-trend of the measured value under the influence of alternating stresses. The decrease in the work function occurs directly above the future crack. The increment in the work function is equal to $-120$ meV after $20 \times 10^6$ cycles.

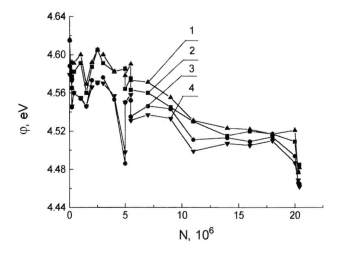

**Fig. 6.6** Work function versus number of cycles for EP866 superalloy. The test frequency $\omega$ equals to 2338 rad s$^{-1}$. Curves 1 - 4 correspond to four different points at the area of the future crack. A maximum in the work function is observed at the curves after $2.5 \times 10^6$ cycles (12.5% of lifetime). It is clear that crystal lattice defects annihilate temporarily.

In the early stages of fatigue tests the increment in the work function is found to have cyclic variations with $N$ (Figure 6.7) as if $\varphi$ is reversible. One can see that every maximum in the work function dependence on the cycle number is less than the former one. There are grounds to believe that up to a certain time the accumulation of defects in the crystal lattice and the relaxation process alternate with each other.

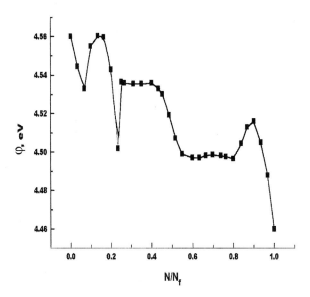

**Fig. 6.7** The effect of fatigue tests upon the work function for EP479 superalloy. $N_f$ is the number of cycles to fracture. $\tau_m$=535 MPa; $\omega$=2338 rad s$^{-1}$. One can see the alternation of an accumulation of crystal lattice defects (minima) and an annihilation of defects (relaxation: maxima).

## 6.2 Dislocation Density in Fatigue-Tested Metals

The results that have been obtained make it possible to assume the existence of two stages of structural changes in fatigue-tested metals:
1. The stage of reversible structural alterations when the work function at a given point of the specimen decreases and increases periodically.
2. The stage of irreversible changes in the surface layer when the work function decreases monotonically owing to the generation of charged surface steps.

These data are of practical significance. Fracture, as a result of fatigue, can possibly be prevented by relaxation processes.

Having the experimental data in mind, we can consider the model of fatigue process and evaluate some physical parameters.

If a metal specimen is subjected to alternating stresses the generation of dislocations is known to occur. There is a threshold stress for this process to begin

$$\tau_s = \frac{\mu_s b (n_{pu}\rho)^{\frac{1}{2}}}{2\pi}, \qquad (6.1)$$

where $\mu_s$ is the shear modulus; $b$ is the Burgers vector; $n_{pu}$ is the number of dislocations in the pile-up; $\rho$ is the initial dislocation density.

The generation of dislocations near to the surface, and their emergence on the surface results in the formation of steps. These steps are known to have dipole moments. Negative charges accumulate between surface atoms. The contribution of dipoles tends to decrease the work function.

The rate of the dislocation multiplication is given [51, 52] by the equation

$$\frac{d\rho}{dt} = \delta \cdot \rho \cdot V, \qquad (6.2)$$

where $\delta$ is a factor of dislocation multiplication; $\rho$ is the dislocation density; V is the dislocation velocity. The velocity of dislocations may be expressed as the velocity of a thermoactivated process:

$$V = V_0 \exp\left[-\frac{U_0 - \gamma(\tau - \tau_s)}{kT}\right] \qquad (6.3)$$

where $V_0$ is a pre-exponential multiplier; $U_0$ is the activation energy for the dislocation motion; $\gamma$ is the activation volume; $\tau$ is the applied shear stress; $k$ is the Boltzmann constant and $T$ is the temperature.

The alternating stress is dependent on time:

$$\tau = \tau_m |\sin \omega t| \qquad (6.4)$$

where $\tau_m$ is the amplitude of the applied stress; $\omega$ is the angular frequency of the stress; $t$ is time. Thus, we obtain

$$t_{s0} = \frac{1}{\omega} \arcsin\left(\frac{\tau_s}{\tau_m}\right); \quad t_{s0} = \frac{\pi}{\omega} - t_{s0} \qquad (6.5)$$

where $t_{s0}$ and $t_{f0}$ are the start and the finish times of the dislocation motion within one half-cycle, respectively. The dislocation density increases from $\rho_0$ to $\rho$ under the effect of alternating stresses. Substituting (6.3) into (6.2) and taking the definite integral we obtain

$$\ln \frac{\rho}{\rho_0} = \delta V_0 \exp\left[-\frac{U_0 + \gamma \tau_s}{kT}\right] \times \int_{t_{s0}}^{t_{sf}} \exp\left[-\frac{\gamma \tau_m |\sin(\omega t)|}{kT}\right] dt \qquad (6.6)$$

We solved (6.6) numerically. The dislocation density in the surface layer was calculated for every cycle. The dislocation density was also determined experimentally by the X-ray method (see Section 2.1.4).

**Table 6.1** Some initial parameters for calculation of the dislocation density.

| Metal | $b$ (nm) | $\mu_s$ (MPa) | $\omega$ (rad s$^{-1}$) | $\tau_m$ (MPa) | $\rho_0$ ($10^{11}$ m$^{-2}$) | n |
|---|---|---|---|---|---|---|
| Al | 0.286 | $2.70 \times 10^4$ | 5880 | 82 | 3.8 | 100 |
| Alloy VT 3-1 | 0.295 | $2.95 \times 10^4$ | 400 | 60–140 | 8.0 | 100 |

The initial values for the calculation of dislocation density during fatigue tests are presented in Table 6.1. The initial rate of dislocation multiplication was taken from experimental data [53].

The results of calculations of the dislocation density in aluminum are compared with experimental data in Figure 6.8. The fit between experimental points and the calculated curve is satisfactory. The increase in the dislocation density is found to be in agreement with the decrease in the work function.

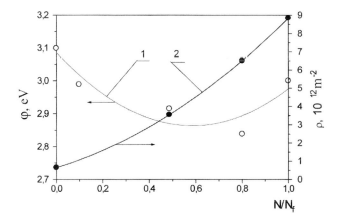

**Fig. 6.8** The effect of fatigue tests on the dislocation density and on the work function for aluminum. Symbols ○ and the curve 1 are values of the measured work function versus the ratio $N/N_f$, $N_f$ is the number of cycles until fracture; 2, the mobile dislocation density, results of calculation; ●, the dislocation density, experimental data.

Results for the titanium-based alloy are presented in Figure 6.9. One can see that curves of $\Delta\varphi$ and $\rho$ appear to be a mirror reflection of each other. The increase in dislocation density is shown to be in agreement with the decrease in the work function. The experimental data on $\rho$ values fit the calculated curves reasonably well.

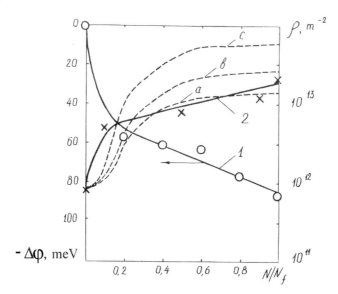

**Fig. 6.9** The effect of fatigue tests on the dislocation density and the work function for the titanium-based VT 3-1 alloy. The symbols ○ and curve 1 are values of the measured work function versus the ratio $N/N_f$; 2, the mobile dislocation density, results of X-ray measurements; a, computed curve according to (6.6), $\tau_m = 60$ MPa; b, as a, 80 MPa; c as a, 120 MPa.

## 6.3
## Summary

The electronic work function has been measured during fatigue tests for aluminum, nickel-based and titanium-based alloys. The work function was found to decrease at the very place where the crack would be formed, long before its creation.

The decrease in the work function under the effect of alternating stresses was also observed on the surface opposite to the surface where a crack is being formed. The observed drop in the work function is not caused by proper crack formation. It reflects structural changes in the specimen surface during stress cycling.

The work function was found to be structure-sensitive. It begins to fall just after 5–10% of the total lifetime of aluminum specimens.

The decrease in the work function was approximately equal to $-120$, $-130$ and $-40$ mV after $10^6$ cycles for EP866, EP479 superalloys, and a titanium-based alloy, respectively.

In the early stages of fatigue tests the increment in the work function was found to have cyclic variations with $N$ as if it is reversible. One could observe an alternation of the crystal lattice defect accumulations (minima at curves

$\varphi - N$) and of the defect annihilations and relaxation (maxima at curves $\varphi - N$). Every maximum at the work function curve was less than the former one. Defect accumulation and defect annihilation (relaxation) were observed to alternate with each other.

The method of work function measurement may be used to predict the initiation of fatigue cracks.

The generation of dislocations near to the surface, and also their emergence results in formation of steps on the surface. These steps have dipole moments. Negative charges accumulate between surface atoms. The contribution of dipoles tends to decrease the work function.

An equation was developed for estimation of the increase in the dislocation density during cycling. The equation was solved numerically. The fit between experimental points and calculated curves was satisfactory. The increase in the dislocation density was found to be in agreement with the decrease in the work function.

# 7
# Computer Simulation of Parameter Evolutions during Fatigue

In the previous chapter experimental data were presented on the change in the structure and work function under the influence of alternating stresses. The next step is to develop a physical model that describes processes which lead to fatigue. Our approach is to work out the model as a system of ordinary differential equations that correspond to parameter changes. We solve the system numerically.

## 7.1
### Parameters of the Physical Model

Five values are of interest for a complete description of the physical mechanism under consideration:

- the threshold stress of the dislocation motion $\tau_s$
- the density of mobile dislocations $\rho$
- the velocity of mobile dislocations $V$
- the density of surface steps $n$
- the electron work function $\varphi$.

All these values depend on time $t$.

## 7.2
### Equations

To obtain a system of ordinary differential equations we should derive formulas that describe the changes in each parameter as a function of both time and other parameters.

## 7.2.1
### Threshold Stress and Dislocation Density

If a metal specimen is subjected to alternating stresses the generation of dislocations is known to occur. There is a threshold stress $\tau_s$ at which this process begins [21]

$$\tau_s = \frac{\mu_s b (n_{pu}\rho)^{\frac{1}{2}}}{2\pi} \tag{7.1}$$

where $\mu_s$ is the shear modulus; $b$ is the Burgers vector; $n_{pu}$ is the number of dislocations in the pile-up; $\rho$ is the density of dislocations. In metals and alloys the initial value of $\rho$ is of the order of $10^{10} - 10^{12}$ m$^{-2}$. The dislocation density increases during the deformation.

Differentiating (7.1) with respect to $t$ we arrive at

$$\frac{d\tau_s}{dt} = \frac{\mu_s b}{4\pi} \cdot \left(\frac{n_{pu}}{\rho}\right)^{\frac{1}{2}} \cdot \frac{d\rho}{dt} \tag{7.2}$$

The number of newly generated dislocation loops is directly proportional to the mobile dislocation density and also to the dislocation velocity [51, 52]. Hence the multiplication rate of mobile slip dislocations is given by

$$\frac{d\rho}{dt} = \delta \rho V \tag{7.3}$$

where $\delta$ is a coefficient of multiplication of mobile dislocations. The coefficient $\delta$ has the unit of inverse length.

Substituting (7.3) to (7.2) we obtain the first differential equation of the system:

$$\frac{d\tau_s}{dt} = \frac{\delta \mu_s b V (n_{pu}\rho)^{\frac{1}{2}}}{4\pi} \tag{7.4}$$

The threshold stress increases with time.

Equation (7.3) is the second equation of the desired system.

## 7.2.2
### Dislocation Velocity

Taking the derivative of (6.3) and also taking into account (6.4) one obtains

$$\frac{dV}{dt} = \frac{V_0 \gamma}{kT} \exp\left[-\frac{U_0 - \gamma(\tau_m |\sin \omega t| - \tau_s)}{kT}\right] \left[\omega \tau_m |\cos \omega t| + \frac{\delta \mu_s b V (n_{pu}\rho)^{\frac{1}{2}}}{4\pi}\right]$$

$$\tag{7.5}$$

The pre-exponential factor, $V_0$, is estimated as [21]

$$V_0 = \frac{\nu_D b^2}{l} \tag{7.6}$$

where $\nu_D = 10^{13}$ s$^{-1}$ is the Debye frequency, $l = 1/\sqrt{\rho} \approx 10^{-6}$ m is the mean length of the dislocation segments. Thus, $V_0 \approx 0.818$ m s$^{-1}$ for aluminum.

The activation volume $\gamma = b^3$. The activation energy $U_0$ of the dislocation slip for aluminum can be estimated approximately [21] as 0.16 eV at$^{-1}$ =2.56 × $10^{-20}$ J at$^{-1}$.

### 7.2.3
### Density of Surface Steps

Dislocations that emerge on the surface cause the creation of steps. The equation of conservation of the crystal lattice defects can be written as

$$(C\rho)^{\frac{1}{2}} \cdot L \cdot V \cdot dt = dn \cdot h \cdot L \tag{7.7}$$

where $\rho$ is the mobile dislocation density as before; $L$ is the size of the crystal; $Vdt$ is the mean path of dislocations to the surface; $dn$ is the increment in the surface step density; $h$ is the mean height of the steps. One usually observes monoatomic steps [26], hence $h = b$. The coefficient $C < 1$ reflects the fact that only part of mobile dislocations can emerge on the surface and create steps.

Thus, the rate of formation of surface steps is related to the mobile dislocation density and to the velocity of their motion:

$$\frac{dn}{dt} = \frac{V(C\rho)^{\frac{1}{2}}}{h} \tag{7.8}$$

This is our third equation.

### 7.2.4
### Change in the Electronic Work Function

The increment in the work function $\varphi$ due to an increase in the step density is given by [26]

$$\Delta \varphi = -A\mu n \tag{7.9}$$

where $\varphi$ is in eV; $A$ is a constant which is equal to $3.77 \times 10^{-19}$ eV · m$^2$ D$^{-1}$; $D$ is the Debye, that is, the unit of the dipole moment, $D = 3.3356 \times 10^{-30}$ C · m; $n$ is the step density in m$^{-1}$; $\mu$ is the dipole moment divided by the interatomic distance in D m$^{-1}$. $\mu = e \times b/b \times 3.34 \times 10^{-28} = 4.79 \times 10^{-10}$ D m$^{-1}$.

Differentiating (7.9) we arrive at

$$\frac{d\varphi}{dt} = -\frac{A\mu V(C\rho)^{\frac{1}{2}}}{h} \tag{7.10}$$

## 7.3
## System of Differential Equations

We have obtained the system of five ordinary differential equations consisting of (7.3), (7.4), (7.5), (7.8), (7.10). The equations are in canonical form. This is the system to be used for computer simulation.

The general form of a set of $N$ first-order differential equations for the unknown functions $y_i$, $i = 1, 2, \ldots, y_N$ is

$$\frac{dy_i}{dt} = f_i(t, y_1, y_2, \ldots, y_N) \tag{7.11}$$

where the functions $f_i$ on the right-hand side are known. In our case $N = 5$.

We use the Runge–Kutta method [54] for integration of differential equations. Runge–Kutta methods propagate a numerical solution over the time interval by combining the information from several Euler-style steps (each involving one evaluation of the right-hand side of the equations) and then using the information obtained to match a Taylor series expansion up to some high order. Program MATLAB enables one to solve the system and achieve a specified precision of fourth order. We use the so-called ODE45 Runge–Kutta method with a variable step size. The step size is continually adjusted to achieve a specified precision.

## 7.4
## Results of the Simulation: Changes in the Parameters

Our computer model is semi-empirical. Equations of the system contain only values which have a certain physical meaning. We should make some reasonable assumptions because some values are unknown.

The following initial values for the parameters are chosen:

| | |
|---|---|
| time | $t = 0$ |
| threshold stress | $\tau_s = 37.6$ MPa |
| dislocation density | $\rho = 2 \times 10^{12}$ m$^{-2}$ |
| dislocation velocity | $V = 1.78 \times 10^{-8}$ m s$^{-1}$ |
| surface step density | $n = 0$ |
| electronic work function | $\varphi = 3.1$ eV |

Initial values of $\tau_s$, $V$ are calculated by corresponding formulas. $n_{pu}$ is assumed to be equal to 100; the coefficient of dislocation multiplication $\delta = 2 \times 10^4$ m$^{-1}$ [55]; multiplier $C = 1 \times 10^{-7}$; the number of cycles $N = 1.87 \times 10^6$; the test time equals 2000 s.

In Figure 7.1 the data obtained from the model are presented for fatigue-tested aluminum.

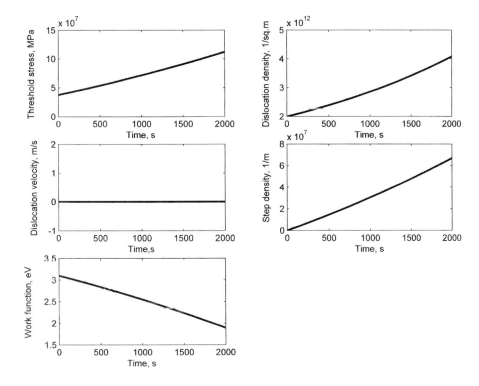

**Fig. 7.1** Dependence of the structure parameters and the work function on time of cycling for aluminum. The computer simulation of the fatigue test by a system of five ordinary differential equations. The stress amplitude $\tau_m = 82$ MPa, the angle frequency $\omega = 5881$ rad s$^{-1}$.

The density of dislocations increases during the test from $2.0 \times 10^{12}$ to $4.1 \times 10^{12}$ m$^{-2}$. The threshold stress increases correspondingly. The dislocation velocity is almost constant on the shown scale. Steps appear on the surface of the metal. The concentration of steps increases up to $7 \times 10^7$ m$^{-1}$. The mean calculated distance between neighboring steps is 14.3 nm or $\approx 49.5b$. The electron work function decreases almost linearly from 3.1 (an initial experimental value) to 2.0 eV.

In Figure 7.2 the model data are presented for the EP866 superalloy. The test time is equal to 3000 s. The density of surface steps increases up to $4 \times 10^7$ m$^{-1}$ for the nickel-based superalloy. The work function decreases by $-600$ meV.

It might seem that the coefficient of dislocation multiplication $\delta$ has been chosen somewhat arbitrarily. We are forced to consider it as a fitting coefficient because its value is unknown. However, it turned out that a change in the value of $\delta$ has only a small effect on the results of the calculations.

# 7 Computer Simulation of Parameter Evolutions during Fatigue

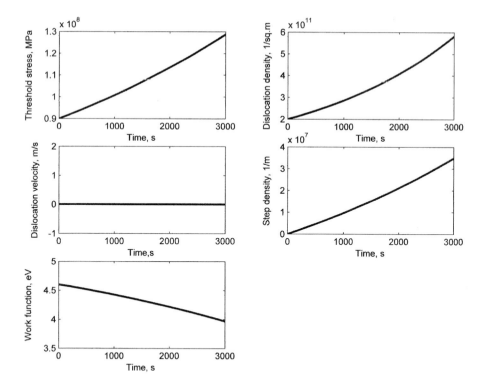

**Fig. 7.2** Dependence of the structure parameters and the work function on time for the EP866 superalloy. The computer simulation of the fatigue test uses a set of five ordinary differential equations. The stress amplitude $\tau_m = 530$ MPa, the angle frequency $\omega = 2338$ rad s$^{-1}$ (see text for details).

The authors of [26] have measured the decrease in the work function as $(-200 - -300)$ meV for gold and platinum with the increase in the step density up to $(1.5 - 6.0) \times 10^8$ m$^{-1}$, see Figure 3.9.

One can see that the results obtained using the physical model (Figures 7.1 and 7.2) and the experimental data correspond well with each other.

## 7.5
## Summary

A system of differential equations has been proposed to simulate the evolution of parameters during fatigue. The evolution of five values was studied: the threshold stress; the density of mobile dislocations; the velocity of mobile dislocations; the density of surface steps and the electronic work function.

Formulas that describe the changes in each parameter as a function of time and other parameters have been derived; a system of five ordinary differential equations was obtained. Equations of the system contain only values which have a certain physical meaning. The Runge–Kutta method was used for integration of the system.

Results following out the physical model fit the experimental data well.

The density of dislocations increased by twice as much for fatigued aluminum and three times for the fatigued superalloy. The threshold stress increased correspondingly.

Steps appeared on the metallic surfaces. The concentration of steps increased to $7 \times 10^7$ for aluminum and to $4 \times 10^7$ m$^{-2}$ for the nickel-based superalloy. That is close to experimental values. The electron work function decreased linearly as in reality.

# 8
# Stressed Surfaces in the Gas-Turbine Engine Components

Modern aviation gas-turbine engines have to be more efficient and reliable. The components of these engines must provide fatigue strength, structural stability and excellent creep properties. The failure of a gas-turbine blade is always catastrophic.

It is known that, in order to obtain maximum performance from a specific gas turbine, it is necessary for the temperature of the gas to be as high as possible. However, the maximum temperature which can be reached in use, is limited by the properties of the materials.

The prevention of failure due to the high-cycle fatigue in gas-turbine engine components is one of the most critical challenges facing the aircraft industry. The gas-turbine blades and disks are also subjected to a low-cycle fatigue associated with take-off and landing. An improvement in fatigue life can be achieved by introducing favorable surface residual stresses.

In this chapter we consider the distribution of induced residual stresses for real engine parts and the effect of various surface treatments on fatigue strength.

## 8.1
### Residual Stresses in the Surface of Blades and Disks and Fatigue Strength

The rotor blades in gas turbines do not form a single body with the rotor disk, but are retained by means of their base extensions in appropriate seats provided on the circumference of the disk. These seats have sides with a grooved profile, in which the end portion of the root of the corresponding blade is engaged.

The surface of loaded components of gas turbine engines is treated in a special way in order to insure a sufficient endurance strength and to prevent fatigue failure. One way of preventing fatigue fracture is to reduce or eliminate the tensile stresses that occur during oscillating loading. Favorable compressive residual stresses are induced by different surface treatment processes.

*Strained Metallic Surfaces.* Valim Levitin and Stephan Loskutov
Copyright © 2009 WILEY-VCH Verlag GmbH & Co. KGaA, Weinheim
ISBN: 978-3-527-32344-9

### 8.1.1
### Turbine and Compressor Blades

Turbine blades operate at high temperatures within an aggressive environment. In addition they undergo thermal cycles as aircrafts take-off and land. Cases of root fracture and break-away of a blade from the disk are especially dangerous. That is why the reliability of the connection of the blade to the disk is so important.

One a problem which is particularly significant is that of guaranteeing an optimal connection of the blades to the rotor disk, in all conditions under which the machine will functioning.

It should be noted that the method of connection of the blades on the rotor disk represents a crucial aspect of the design of any rotor, taking into account the fact that the disk must withstand loads generated by the blades without giving rise to origination of cracks and breakages.

During the functioning of the machine, the rotor blades are subjected to high stresses both in the radial direction and, to a lesser extent, in the axial direction. The radial stresses are caused by the high angular speed of the turbine, whereas the axial stresses are caused by the effect produced by the flow of gas on the profiled surfaces of the blades. The most widely used connection at present is commonly known as the pine tree type, so-named because of the shape of the blade root, which in its cross-section resembles an upturned pine tree.

These roots are connected to seats or coupling slots complementary to them, which are provided on the circumference of the rotor disk, such that the grooves in the sides of the seat correspond to the teeth of the root and a groove at the base of the seat corresponds to the lower end of the root.

This type of connection has areas of particular stress concentration which can be determined more specifically as being at the bottom of the groove, on the base of the seat, and on the base of the grooves of each tooth, which constitutes the actual connection profile.

We have studied the effect of strengthening conditions on the sign, value and distribution of residual stresses over the surface of the aircraft turbine blades. The fatigue limit $\sigma_{fl}$ based on $1 \times 10^7$ cycles was also measured for the components[1]. The fatigue limit is defined as the stress amplitude which the specimen or the component can withstand for at least a given number of fatigue cycles.

The manufacture and treatment of real turbine and compressor blades was carried out at a plant in the working environment. The blades of a gas-turbine of high pressure were investigated. The technique of the residual stress determination is described in Section 2.1.3. The turbine blades were manufactured

---

1) The fatigue limit is also termed the endurance limit $\sigma_e$.

of the cast ZhS6K superalloy. The scheme of X-ray measurements of the residual stresses is presented in Figure 8.1. X-ray diffractograms were taken from the shelves, the teeth, and from the tails of blades.

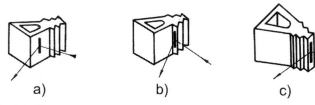

a)  b)  c)

**Fig. 8.1** The scheme of the X-ray study of residual stresses in roots of the cooling turbine blades: a, point of investigation is the shelf; b, the diffractogram is taken from the first tooth of the root; c, the diffractogram is taken from the tail. Incident and reflected rays are shown.

The strengthening treatment of blades is carried out by means of a special installation with ball-bearings as a working tool. Balls of diameter from 0.4 to 3.0 mm are put in motion by chamber walls vibrating from a generator of ultrasonic oscillations with a frequency of 17.2 kHz and the amplitude of 45 μm. This treatment is supposed [56] to be a relatively light of treatment because of the low energy of the balls under multiple blows to the surface. The method allows one to strengthen thin-walled components. The size of the working balls, their total mass in the chamber and the amplitude of the chamber-wall oscillations can be changed to optimize the results of the treatment.

One can see the experimental data in Table 8.1. The surface residual macroscopic stresses after the treatment are from $-488$ to $-702$ MPa. The minus means that profitable compressive stresses are induced. However, the first tooth of the blade remains unstrengthened ($\sigma = -16$ MPa). It is certain that the first tooth is situated in a shadow of the field of vibrating ball-bearings.

**Table 8.1** Residual stresses in the root surface of turbine blades (MPa).

| Point | Before strengthening | After strengthening | After fatigue tests |
|---|---|---|---|
| Shelf | $-56$ | $-702$ | $-685$ |
| The first tooth | 0 | $-16$ | $-49$ |
| Tail | 0 | $-498$ | $-488$ |

So, one should change the position of the complex shaped component in the working chamber during the procedure in order to ensure the homogeneity of the surface treatment of the component.

Further batches of workpieces were treated under four machining conditions (denoted as Y1, Y2, Y3, Y4). We measured the residual stresses at two points for the blade shelves and at four points for the teeth. The data obtained are presented in Table 8.2.

## 8 Stressed Surfaces in the Gas-Turbine Engine Components

**Table 8.2** Effect of surface treatment on residual stresses $\sigma$ and the fatigue limit $\sigma_{fl}$ of gas-turbine blades.

| Procedure conditions | Without treatment | | Y1 | | Y2 | | Y3 | | Y4 | |
|---|---|---|---|---|---|---|---|---|---|---|
| Amplitude (μm) | – | | 100–140 | | 80–100 | | 80–100 | | 80–100 | |
| Diameter of balls (mm) | – | | 0.9–1.3 | | 1.0–1.3 | | 1.0–1.3 | | 1.0–1.3 | |
| Mass of balls (kg) | – | | 0.175 | | 0.050 | | 0.030 | | 0.030 | |
| Time (s) | – | | 60 | | 45 | | 65 | | 90 | |
| Number of the blade | N832 | N851 | E902 | I993 | E59 | O21 | E78 | O764 | E894 | E903 |
| $\sigma$ on the shelf (MPa) | 0 | 0 | –524 | –466 | –510 | –486 | –584 | –558 | –554 | –557 |
| $\sigma$ on the first tooth (MPa) | +20 | +121 | –229 | –46 | –130 | –168 | –332 | –230 | –333 | –184 |
|  | +59 | +65 | –230 | –340 | –342 | –57 | –163 | –170 | –138 | +34 |
| $\sigma_{fl}$ (MPa) | – | | 117.6 | | 126.8 | | 126.8 | | 117.6 | |

Just-manufactured blades have no residual stresses at all or have small tensile stresses on their teeth (see the second column in Table 8.2). The necessity for the surface strengthening of the blade roots is obvious. It is easy to induce the favorable stresses on the plane surface of shelves (see the sixth line of Table 8.2). The compressive residual stresses from –466 to –584 MPa are induced in the shelves as a consequence of the treatment of the blade root surfaces.

The level of compression residual stresses on teeth is usually lower than that of on shelves. The distribution of stresses is not uniform. For example, the residual stresses vary from –46 to –340 MPa for regime Y1. This nonuniformity creates a dangerous gradient of stresses in the root surface.

The treatment regime denoted by Y3 turned out to be favorable. The best performance includes a treatment time of 65 s, a total ball mass of 0.030 kg, the diameter of the balls from 1.0 to 1.3 mm. The induced stresses on the first tooth vary from –163 to –332 MPa. The fatigue limit of turbine blades on a basis of 10 millions cycles is as high as 126.8 MPa.

Consequently, X-ray monitoring of the surface treatment is required.

### 8.1.2
### Grooves of Disks

The scheme of the X-ray investigation of disks is shown in Figure 8.2.

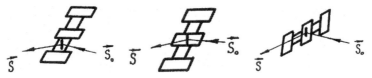

**Fig. 8.2** The scheme of the X-ray study of residual stresses in the root parts of a disk. $\vec{s_0}$ and $\vec{s}$ are incident and reflected X-rays, respectively.

The state of the surface of these essential gas turbine components depends substantially upon a manufacturing technique. Grooves of disks are made by the broaching.

Data on the X-ray study of disk grooves are illustrated in Figure 8.3. For the surface before strengthening the curve $2\theta - \sin^2 \psi$ denoted as B is parallel to the axis of abscissas. Hence the residual stresses are zero. After the surface treatment of the disks the dependences are straight lines with a positive slope to the axis of abscissas (Figure 8.3, curves C and D). Thus, $\partial(2\theta)/\partial(\sin^2 \psi) > 0$. According to (2.13) it means that the surface residual stresses are negative and therefore they are compressive.

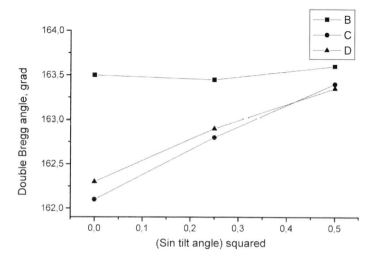

**Fig. 8.3** The double Bragg angle $2\theta$ versus $\sin^2 \psi$, where $\psi$ is the angle between the normal to the surface and the normal to the reflected crystal plane. B, the curve for a groove of a just-manufactured gas-turbine disk; curves C and D, as B, after surface treatment by ball-bearings oscillating in the ultrasonic field.

In Table 8.3 the data on residual stresses after the broaching of grooves are presented. As a result of mechanical processing the surface of the grooves acquires nonuniform or tensile stresses. The subsequent treatment is of fundamental importance. As one can see from Table 8.3 the strengthening induces favorable compressing stresses.

Fatigue tests of disks confirm efficiency of the surface treatment for high-performance applications. The dependence of the number of cycles until fracture on the time of treatment is shown in Figure 8.4. One can see that the lifetime of disks can be increased by as much as three times if one applies ball-bearings balls 1.0 mm diameter as working tools.

**Table 8.3** Residual stresses $\sigma$ in the surface of disk grooves after manufacturing by the broaching.

| Alloy | No | Procedure conditions | $\sigma$ (MPa) |
|---|---|---|---|
| EI698 | 1 | Broaching at a speed of 1 m min$^{-1}$ | Nonuniform |
|  | 2 | Broaching at a speed of 4 m min$^{-1}$ | +660 |
|  | 3 | As 1 and strengthening by ball-bearings in ultrasonic field | −430 |
|  | 4 | As 2 and strengthening by ball-bearings in ultrasonic field | −600 |
| VT 9 | 5 | Broaching at a speed of 8 m min$^{-1}$ | Nonuniform |
|  | 6 | As 5 and strengthening by ball-bearings in ultrasonic field | −700 |

The application of balls of 0.68 mm diameter does not result in a change in the durability of disks. It is clear that an accurate procedure should be employed and controlled for high-performance application.

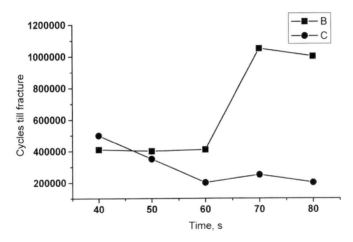

**Fig. 8.4** The effect of the surface strengthening by oscillating ball-bearings in the ultrasonic field on the lifetime of disk slots. B, diameter of ball-bearings is 1.0 mm; C, diameter of balls is 0.68 mm.

## 8.2
### Compressor Blades of Titanium-Based Alloys

The blades of the aircraft gas-turbine compressor are subjected to alternating-sign stresses and centrifugal forces. Failure analysis reveals that ruptures of blades during operation can occur because of technological surface stress concentrators[2]. Ensuring fatigue life and the reliability of blades is based on the

2) Or owing to collisions with foreign bodies during operation.

careful removal of surface defects like hairlines, tearing and chips. Keeping this in mind, one subjects the surface of the blades to a finishing strengthening procedure and to close control.

A blade is shown in Figure 8.5. The nominal dimensions of a blade are presented in Figure 8.6. The strengthening of blades is a complicated problem since the components have feathered edges of up to 0.5 mm thick, and rounded radii of up to 0.2 mm (see Figure 8.6). There is the risk of a surface over-hardening and of damage to the feathered edges. For these reasons the problem of blade strengthening has not yet been solved sufficiently.

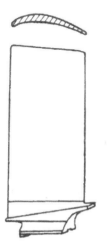

**Fig. 8.5** A blade of the gas-turbine compressor.

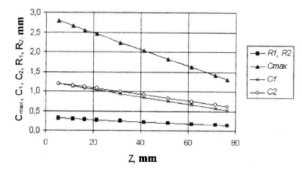

**Fig. 8.6** Dimensions of compressor blades: $R_1$ and $R_2$ are radii of rounding of edges; $C_{max}$ is the maximum thickness; $C_1$ is the dimension of the input edge; $C_2$ is the dimension of the output edge.

Compressor blades have, for many years, been manufactured from titanium-based alloys. For these blades the conventional fine-grain alloys of systems Ti-Al-Mo and Ti-Al-Mo-Cr (Table 2.2) are used.

# 8 Stressed Surfaces in the Gas-Turbine Engine Components

The diffractogram of the titanium-based VT 8 alloy is illustrated in Figure 8.7. The structure of the alloy consists mainly of an α-phase that has a hexagonal crystal lattice. The second structural component is the β-phase, which has a cubic crystal lattice.

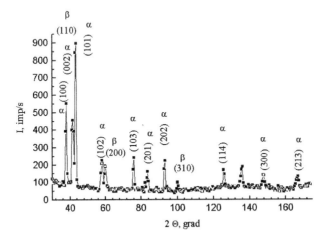

**Fig. 8.7** The X-ray diffractogram of the VT 8 alloy. The monochromatic β-radiation of the cobalt X-ray tube, $\lambda = 0.162075$ nm. The Miller indexes (hkl) of the reflecting crystal planes of the α and β phases are indicated.

The lattice parameters of the phases are as follows: $a = 0.2953$ nm, $c = 0.4729$ nm for α-Ti; $a = 0.3320$ nm for β-Ti. The structure of VT 3-1 and VT 8 alloys is characterized by metastability. The transformations $\beta \rightarrow \omega$; $\beta + \omega \rightarrow \alpha + \beta$; $\beta \rightarrow \alpha'$, and others, are known to occur in microscopic volumes. The composition of alloys and also heat treatment will influence these transformations which result in changes in the intensity of X-ray reflections.

We have studied batches of industrial blades made of the titanium-based alloy VT 8. The vibration polishing of blades, shot peening and treatment by ball-bearings in an ultrasonic field were applied as comparative technologies.

## 8.2.1
### Residual Stresses and Subgrain Size

Figure 8.8 illustrates the points of the X-ray investigation. The points 1 and 2 are situated on the back of blades; points 3, 4, 5, 6 are located on the edges.

After annealing the blades at 923 K for 3 hours, the residual stresses in the surface are tensile and equal to +15 – +35 MPa. The broadening of the X-ray (213) reflection equals 3.0–3.2°.

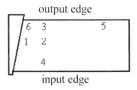

**Fig. 8.8** Points of X-ray investigation on the surface of compressor blades.

The grinding and fine-finishing of blades results in the occurrence of inhomogeneous macroscopic stresses of the order of −400 to −600 MPa. The reflection broadening is 4.5 to 5.0°.

The level of residual stresses for hardening by ball-bearings in the ultrasonic field is shown in Figure 8.9. The treatment induces compressive residual stresses. The back surface of the blades is strengthened considerably more than the surfaces of the thin edges. The scattering of data is relatively high: from −350 to −1000 MPa.

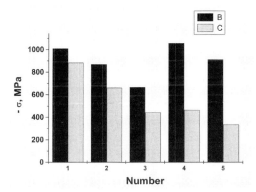

**Fig. 8.9** Residual stresses in the surface of compressor blades after treatment by ball-bearings in the ultrasonic field. Data for five blades. B, the back of the blades; C, the edge of the blades.

The transmission electron microscopy allows one to determine an increase in dislocation density by two or three orders of magnitude after treatment. We observe entangled dislocations near the surface. The depth of strengthening is as high as 300 μm.

The shot peening of surface effects are shown in Figure 8.10. The mean values of the residual stresses are lower by 40% than after treatment in the ultrasonic field. However, the homogeneity of stress distribution is favorable.

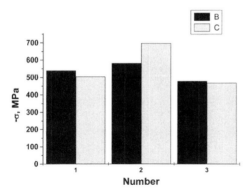

**Fig. 8.10** Residual stresses in the surface of compressor blades after treatment by shot peening. Data for three blades. B, the back of the blades; C, the edge of the blades.

Residual stresses vary between −500 and −700 MPa. Backs and edges are strengthened almost equally. The depth of compressing stresses after shot peening is found to be equal to 250 μm.

The broadening of X-ray reflections is related to an increase in the microscopic stresses and a decrease in the subgrain size. The treatment of blade surfaces by steel balls in the ultrasonic field induces broadening from 5.3 to 6.0° for the blade backs and less (4.4 to 5.6°) for blade edges (Figure 8.11).

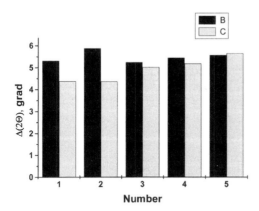

**Fig. 8.11** The broadening of the (213) X-ray reflection for the blade surface after treatment by ball-bearings in the ultrasonic field: B, the back of the blades; C, the edge of the blades.

One can see in Figure 8.12 that shot peening processing results in a more homogeneous distribution of the broadening of the X-ray reflection. Its value is about 5°. We use the broadening of (102) and (213) reflections to determine the microscopic stresses $\Delta a/a$ and the mean subgrain size $D$ for a $\beta$-titanium

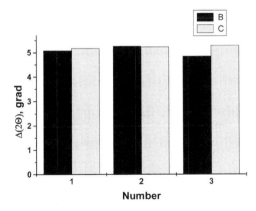

**Fig. 8.12** The broadening of the (213) X-ray reflection for the blade surface after shot peening: B, the back of the blades; C, the edge of the blades.

structure (see Section 2.1.4). Data for the measurement of microscopic stresses and subgrain size in the blade surface are presented in Table 8.4. Microscopic stresses are larger for the thin input and output edges of compressor blades than they are at the back. The subgrain size is found to be on a nanometric scale: from 9.4 to 17.1 nm. We see from Table 8.4 that subgrains decrease in size at the edges of the blades to a greater extent than at the backs. Moreover, the subgrains there have a more homogenous distribution.

**Table 8.4** Microscopic stresses $\Delta a/a$ and subgrain sizes $D$, for compressor blades.

| Process conditions | $D$ (nm) | | $\Delta a/a$ $(10^{-4})$ | |
|---|---|---|---|---|
|  | back | edge | back | edge |
| Annealing at 1173 K (900°C) for 3 hours | > 100 | > 100 | 0 | 0 |
| Strengthening by ball-bearings in ultrasonic field | 14.0 ± 3.4 | 9.4 ± 0.4 | 22.0 ± 3.6 | 27.1 ± 4.1 |
| Shot peening | 17.1 ± 4.5 | 10.6 ± 2.1 | 17.7 ± 4.6 | 30.0 ± 4.2 |

## 8.2.2
### Effect of Surface Treatment on Fatigue Life

Fatigue tests of blades are carried out by means of a vibration table at the resonance eigenfrequency from 400 to 430 Hz. An decrease in the eigenfrequency by 3–5% corresponds to the onset of blade fracture. The tests are performed on the base of $N_b = 10^7$ cycles. The fatigue limit $\sigma_{fl}$ is determined under these conditions. The fatigue test data are illustrated in Table 8.5.

The factor of survivability is presented in the sixth column of Table 8.5. The survivability factor value quantitatively characterizes the fatigue life of a sta-

**Table 8.5** Data of fatigue testing of compressor blades. The material of the blades is the titanium-based VT 8 alloy. Test basis is $10^7$ cycles. $\tau$ is the time of shot peening, $P$ is the pressure of air in pneumatics. $K_i$ is a factor of survivability, $S_\sigma$ is a factor of durability scattering.

| Processing | Label | P (MPa) | $\tau$ (min) | Fatigue limit (MPa) | $K_i$ | $S_\sigma$ ($10^6$ cycles) |
|---|---|---|---|---|---|---|
| Vibratory polishing | V1 | – | – | 545 | 0.066 | 17.2 |
| Treatment in ultrasonic field | U1 | – | – | 575 | 0.127 | 13.8 |
| Shot peening | P1 | 0.08 | 3 | 575 | 0.349 | 16.3 |
|  | P2 | 0.14 | 3 | 575 | 0.120 | 13.9 |
|  | P3 | 0.16 | 3 | 600 | 0.118 | 15.2 |
|  | P4 | 0.08 | 6 | 600 | 0.117 | 15.5 |
|  | P5 | 0.20 | 2 | 595 | 0.115 | 13.6 |
|  | P6 | 0.14 | 4 | 600 | 0.136 | 14.7 |

tistical population of components [57]. The greater this factor the greater is the average life of the population.

It is expressed as

$$K_i = \frac{\sum_{i=1}^{n} \sigma_i N_i}{n N_b \sigma_{fl}} \quad (8.1)$$

where $\sigma_i$ and $N_i$ are the stress amplitude and the number of cycles up to the fracture of the $i$th blade, respectively, $n$ is the number of tested blades, $N_b$ is the test basis, here $10^7$ cycles and $\sigma_{fl}$ is the value of the fatigue limit, prescribed beforehand.

The scattering of the blade durability, $S_\sigma$, is shown in the last column of Table 8.5.

$$S_\sigma = \left[ \frac{\sum_{i=1}^{n} (\overline{N_f} - N_{fi})^2}{n-1} \right] \quad (8.2)$$

where $\overline{N_f}$ is the average number of cycles until fracture, $N_{fi}$ is the number of cycles until fracture for the $i$th blade.

The treatment of the blade surface by ball-bearings in the ultrasonic field increases the fatigue limit by 6% and decreases $S_\sigma$ by 20% as compared with vibration polishing. After shot peening the fatigue limit is increased by 35–55 MPa, whereas the scattering of the blade durability is appreciably decreased.

Processing, labeled as P3 and P4, are considered to be optimal. They cause no damage to the blade feathered edges. This shot peening technology ensures that $\sigma_{fl} = 600$ MPa, $K_i = 0.117$, $S_\sigma = 15.5 \times 10^6$ cycles.

We examined the question of whether the work function can be used to predict fatigue crack origination on the surface of compressor blades. The work function was measured with a step of 1 mm along the back and the trough of blades (that is, the contact potential difference was scanned experimentally). Lines of scanning are shown in Figure 8.13.

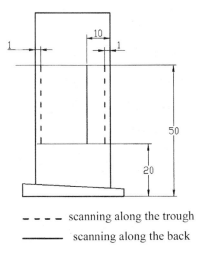

- - - - scanning along the trough
———— scanning along the back

**Fig. 8.13** The scheme of scanning for the blades under study.

The distribution of the surface potential along blades is presented in Figure 8.14. The negative values of the contact potential difference are plotted along the ordinate. According to obtained data the decrease in the work function allows one to predict the place of a future crack. The predicted location of the crack lies near to its actual location. This take place under a stress amplitude of 770 MPa (38 mm from the clamp) and under a stress amplitude of 660 MPa (45 mm from the clamp), (Figure 8.14a,b). A prediction, in another cases, is satisfactory for the output edge or input edge (Figure 8.14c, d).

The fatigue of blades manufactured from VT3-1 alloy after the same treatment was also investigated. The eigenfrequency of oscillations was 320 Hz at the symmetric cycle ($R = -1$). The base of fatigue tests was chosen to be relatively high; $10^8$ cycles. Ten blades were tested for every surface treatment and every stress amplitude of the cycling. We also performed elastic impact tests on the blades by means of an impact machine (Figure 8.15).

The data obtained are presented in Table 8.6. It is clear that shot peening is an optimal variant of blade treatment. Shot peening ensures high values of the surface compressive residual stresses and high mechanical characteristics of the blades.

"The fatigue phenomenon has a long lasting reputation of considerable scatter of fatigue lives in nominally similar fatigue tests" [89]. Figure 8.16 [58]

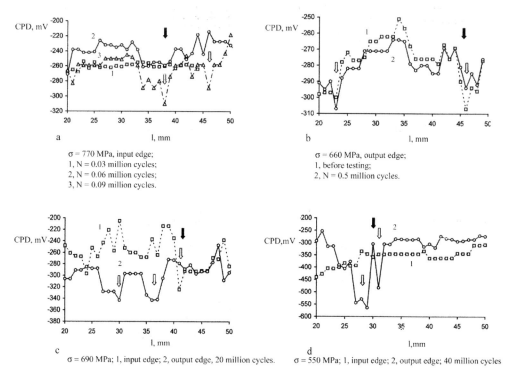

**Fig. 8.14** Distribution of the contact potential difference (CPD) along the surface of blades. Light arrows show a predicted location of the crack; dark arrows show the actual location of the fracture.

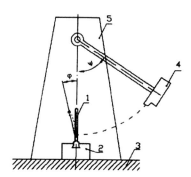

**Fig. 8.15** The impact machine for testing compressor blades: 1, blade; 2, holder; 3, foundation; 4, striker; 5, frame.

presents data of fatigue tests of compressor blades which confirm this opinion.

However the advantage of the shot peening over the surface treatment by bearing balls in the ultrasonic field is evident.

## 8.2 Compressor Blades of Titanium-Based Alloys

**Table 8.6** Data for fatigue testing of compressor blades manufactured from the titanium-based VT3-1 alloy. The base of the tests is $10^8$ cycles.

| Number | Processing | Residual stresses (MPa) | | Fatigue limit (MPa) | Blows up to failure |
|---|---|---|---|---|---|
| | | input edge | back | | |
| 1 | Vibratory polishing | −521 | −531 | 425 | 3 |
| 2 | Shot peening | −605 | −688 | 450 | 4 |
| 3 | Treatment in ultrasonic field | −447 | −457 | 450 | 3 |
| 4 | As 3 and annealing | −38 | −146 | 405 | 4 |

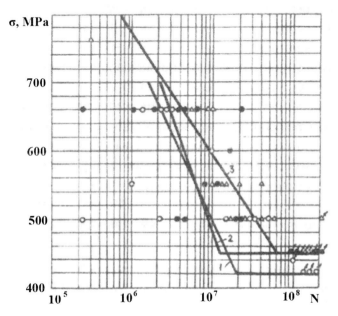

**Fig. 8.16** Data for fatigue tests of compressor blades: 1,∘ vibratory polishing; 2,• as 1 and treatment by ball-bearings in the ultrasonic field; 3,△ as 1 and shot peening. The titanium-based VT3-1 alloy (after [58]).

### 8.2.3
### Distribution of Chemical Elements

Alloys for compressor blades are known to have a two-phase $\alpha + \beta$ structure. The alloy elements are divided as $\beta$-stabilizers (molybdenum, vanadium) and $\alpha$-stabilizers (aluminum, titanium, nitrogen).

We studied the distribution of nitrogen, molybdenum, aluminum and titanium between structural components of titanium-based alloys. The method of X-ray spectrum analysis was applied. The representative data are illustrated in Figures 8.17 and 8.18.

Molybdenum mainly dissolves in the β-phase. Peaks of the molybdenum characteristic radiation $L_{Mo_\alpha}$ correspond to areas of the bright β-phase. On the contrary, nitrogen, titanium and aluminum are displaced in the α-matrix. One can see that reflection minima of these elements correspond to maxima of molybdenum. A surface layer enriched by nitrogen as a result of the high-temperature ion-plasma nitriding is clearly seen.

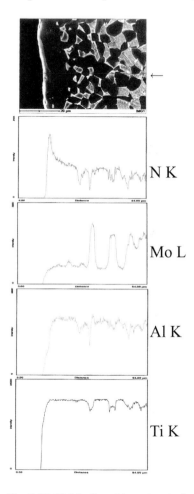

**Fig. 8.17** Distribution of four chemical elements between phases in VT 8 titanium-based alloy. The arrow shows the path of scanning. A bright phase is the high-temperature cubic β-phase, the matrix consists of a hexagonal α-phase.

Nitrogen dissolves in the hexagonal crystal lattice of the α-phase. A strong phase of the nitriding operation also leads to the formation of titanium nitrides TiN and Ti$_2$N. The reflections of these compounds are sufficiently intense to be seen in X-ray diffractograms, Figure 8.19.

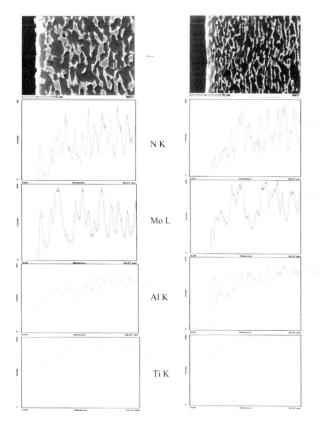

**Fig. 8.18** Distribution of four chemical elements between phases in the VT 8 titanium-based alloy after deep nitriding of the blades. Nitrogen is contained in the $\alpha$-phase as well as molybdenum.

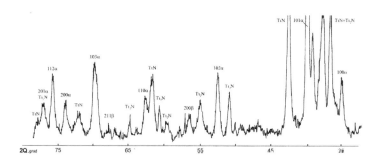

**Fig. 8.19** The X-ray diffractogram of the nitriding layer in the surface of the blade. The $\alpha$-phase, the $\beta$-solid solution and the titanium nitrides TiN and $Ti_2N$ are present in the structure.

## 8.3
## Summary

Modern aviation gas-turbine engines have to meet the higher requirements for efficiency and reliability. Details of these engines must posses fatigue strength, structural stability and excellent creep properties. Blades of the aircraft gas-turbine compressor are subjected during operation to alternating-sign stresses and centrifugal forces. The disk and blades must reliably withstand the loads without giving rise to cracks and breakages.

The seats have sides with a grooved profile, in which the end portion of the root of the corresponding blade is engaged. This type of connection has areas of particular stress concentration.

Ensuring the reliability of blades is based on the careful removal of surface defects like hairlines, tearing and chips. One subjects the surface of the blades to a finishing strengthening procedure to achieve an enhanced fatigue life.

The surface of loaded components of gas turbine engines is usually treated in order to insure a sufficient endurance strength and to prevent the fatigue failure. The method of prevention of fatigue fracture is to eliminate the tensile stresses that occur during fabrication and under the effect of oscillating loading. Favorable compressive residual stresses are induced by different surface treatments.

The strengthening of compressor blades made of titanium-based alloys is a complicated problem since the components have feathered edges, up to 0.5 mm thick and rounded radii up to 0.2 mm. There is the risk of a surface over-hardening and of damage to feathered edges. Owing to these factors, the problem of the blade strengthening has not yet been solved.

The present authors studied the effect of the procedure conditions on the sign, value and distribution of residual stresses over the surface of the aircraft's gas-turbine components. The treatment of the real blades and disks was realized at a plant within the working environment.

Strengthening treatment of the components was carried out specifically by means of the installation with ball-bearings as a working tool. Balls of diameter from 0.4 to 3.0 mm were put in motion by the vibration of chamber walls produced by a generator of ultrasonic oscillations with a frequency of 17.2 kHz and an amplitude up to 140 µm. The other procedures under study were vibratory polishing and shot peening as comparative technologies.

The compressive residual stresses from $-466$ to $-584$ MPa are induced in shelves as a consequence of the treatment of the blade roots of the pine type. The level of compressing residual stresses in teeth is lower than it is in shelves. The distribution of stresses is not uniform. This creates a dangerous gradient of stresses in the root surface. One should change the position of the component in the working chamber during treatment in order to achieve uniformity of treatment.

## 8.3 Summary

Just after mechanical manufacture the surface of the grooves in the disks acquires nonuniform or tensile stresses. The subsequent treatment is of fundamental importance. Strengthening by ball-bearings in the ultrasonic field induces favorable compressing stresses of the order of $-400$ to $-700$ MPa, which can increase the lifetime of disks by as much as three times.

The structure of the titanium-based alloys VT 8 and VT3-1 for compressor blades consists mainly of an $\alpha$-phase that has a hexagonal crystal lattice. The second structural component is the $\beta$-phase which possesses the cubic lattice.

After annealing the compressor blades at 923 K, the residual stresses in the surface are tensile and vary from $+15$ to $+35$ MPa. The broadening of X-ray reflections equals 3.0–3.2°.

The face-grinding and fine-finishing of the blade surfaces results in the occurrence of inhomogeneous residual macroscopic stresses.

The treatment of compressor blades by ball-bearings in the ultrasonic field induces in the surface compressive residual stresses of the order of $-660$ to $-1050$ MPa. The X-ray reflection broadening equals to 5.3–5.6°. The back of blades is strengthened considerably more than the edges. The shot peening processing results in a more homogeneous distribution of microscopic stresses over the blade surface than does treatment by vibrating ball-bearing.

The electron microscopy evidence supports the idea that dislocation density increases by two or three orders of magnitude as a result of the treatment. The depth of strengthening is as high as 300 µm.

Microscopic stresses are larger at the thin input and output edges of compressor blades than they are at the backs. The subgrain size is found to be on the nanometric scale and vary from 9.4 to 17.1 nm. Subgrains decrease in size at the edges of the blades to a greater extent than at the backs. The subgrains there have a more homogenous distribution.

The authors studied the fatigue life of industrial blades made of the titanium-based alloys VT 8 and VT3-1.

The treatment of the blade surface by ball-bearings in the ultrasonic field increased the fatigue limit by 6% and decreased the durability scattering by 20% compared with vibration polishing. After shot peening the fatigue limit increased by 35–55 MPa, up to 600 MPa (the base of tests is $10^7$ cycles), whereas the scattering of the blade durability decreased appreciably.

According to data obtained the decrease in the work function on the blade surface allows one to predict the place of a future crack. The nondestructive prediction of the location of fatigue cracks lies near its actual location.

We studied the distribution of nitrogen, molybdenum, aluminum and titanium between phases of titanium-based alloys. X-ray spectrum analysis was applied. Atoms of molybdenum dissolved mainly in the $\beta$-phase. Nitrogen, titanium and aluminum were displaced mainly in the $\alpha$-matrix.

# 9
# Nanostructuring and Strengthening of Metallic Surfaces. Fatigue Behavior

It is a matter of common agreement that a narrow definition of nanostructures suggests the inclusion objects with at least two dimensions of structure elements below 100 nm ($10^{-7}$ m). Application of nanoterminology to our topic means the size of grains as well as the dimensions of the homogenous stressed areas are of the order of tens of nanometers.

The favorable structural factors that have a significant influence on the fatigue lives and which lead to an additional increase in strength can be enumerated as follows:
- Grains on the nanometric scale.
- Compressing residual macroscopic stresses.
- High dislocation density near the surface.
- Deformation-induced phase transformation.
- Purposeful changes in the chemical composition at a thin surface layer.
- The preferred orientation of grains, that is, a texture.

All these factors, or some of them, are specifically achieved by severe plastic deformation in near-surface regions.

Amongst the various processes of surface treatment are shot peening, treatment by ball-bearings in the ultrasonic field, deep rolling, mechanical attrition, wire-brushing, hammering, and laser shock peening. Such treatments induce structural alterations in the thin surface layer of an alloy. In addition one can strengthen the surface by the method of saturation of surface layers by chemical elements, especially by nitrogen as well as by the coating of components.

One sometimes distinguishes the effect of compressive residual stresses and strain hardening on the fatigue lifetime. In fact, strain hardening should be

**Table 9.1** The effect of structure factors on fatigue crack initiation and crack growth.

| Factors | Crack origination | Crack growth |
|---|---|---|
| Surface roughness and defects | Accelerates | No effect |
| Nanometric scale grains | Retards | Retards |
| High dislocation density | Retards | Accelerates |
| Compressive residual stresses | Retards | Retards drastically |
| Tensile residual stresses | Accelerates | Accelerates drastically |

*Strained Metallic Surfaces.* Valim Levitin and Stephan Loskutov
Copyright © 2009 WILEY-VCH Verlag GmbH & Co. KGaA, Weinheim
ISBN: 978-3-527-32344-9

described in terms of a decrease in the grain size, an increase in the dislocation density, and in a favorable distribution of the crystal lattice defects. The boundaries of grains also have a dislocation character.

Table 9.1 presents the structure factors that considerably affect the initiation and propagation of fatigue cracks.

## 9.1
## Surface Profile and Distribution of Residual Stresses with Depth

The initiation of fatigue cracks is considerable affected by surface roughness.

We have studied compressor blades, the surface of which has been treated

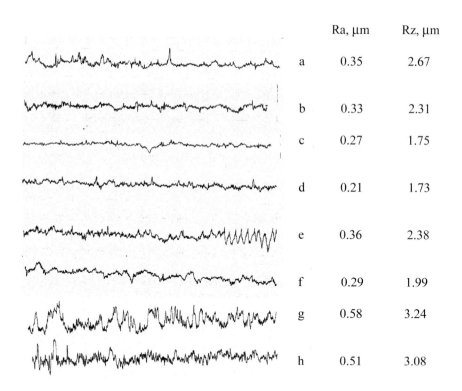

**Fig. 9.1** Surface profiles of compressor blades after different treatments. $Ra$ is the arithmetic average deviation in the profile; $Rz$ is the average height of the peaks. a, vibratory polishing; b, vibratory polishing and ion-plasma nitriding; c, shot peening; d, ion-plasma nitriding and shot pinning; e, strengthening by ball-bearings in the ultrasonic field; f, ion-plasma nitriding and strengthening in the ultrasonic field; g, operating of blades in a gas turbine over 2397 hours; h, as g, operating time of 960 hours.

by means of different techniques. The blades under investigation were first treated by vibratory polishing, plasma nitriding, shot peening or strengthening by ball-bearings in the ultrasonic field. Our aim is to evaluate the effect of the treatments on the surface geometry of the blades and on the distribution of stresses.

The results of measurements of the blade surface profiles are illustrated in Figure 9.1.

The shot peening procedure ensures minimum values of $Ra$ and $Rz$. This is readily seen in profiles $c$ and $d$, Figure 9.1. In this case the average deviation in the profile does not exceed 0.27 µm, the values of $Rz$ are 1.73–1.75 µm.

The operation of blades in a gas turbine leads to an appreciable deterioration in the surface. It is easy to see that deviations in the profiles are double those before the operation (curves $g$ and $h$ in Figure 9.1).

The distribution of residual stresses near the surface of polished blades is shown in Figure 9.2. Vibration polishing is often applied as a finishing operation. One can see (Figure 9.2a) that the compressive (negative) favorable residual stresses from $-300$ to $-400$ are induced near the surface. However, compressive stresses vanish at a depth of 30 µm.

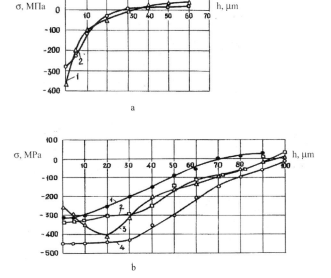

**Fig. 9.2** Distribution of residual stresses with depth for treated blades of the titanium-based VT 8 alloy. a, vibration polishing of the surface: 1, the input edge of the blade; 2, the output edge of the blade. b, shot peening of the surface; process conditions: 1, the pressure, $P = 0.08$ MPa, the time, $t = 360$ s; 2, $P = 0.10$ MPa, $t = 360$ s; 3, $P = 0.08$ MPa, $t = 540$ s; 4, $P = 0.12$ MPa, $t = 540$ s.

Figure 9.2b illustrates effect of the shot peening on the depth and value of residual stresses. Shot peening causes a deeper stress penetration than vibration polishing. A relatively severe condition of the process (a pressure of 0.12 MPa and a treatment time of 540 s) ensures a stress of $-450$ MPa on the surface and a depth of the residual stresses up to 100 μm (curve 4 Figure 9.2). The minimum at $\sigma - h$ curves is typical for stress distribution (curve 3 Figure 9.2).

One applies the ion-plasma nitriding process in order to improve the fatigue lives and erosion-preventive resistance of blades. Figure 9.3a illustrates the stressed state of a surface after ion-plasma nitriding. One can see that only weak positive stresses of the order of 20 MPa remain near the surface. On the surface, the stresses approach zero. As one should expect the residual stresses are negligible since the nitriding process is carried out at a temperature of 773 K (500°C). It is obvious that a strengthening of the surface is necessary after nitriding. It can be easily seen in Figure 9.3b that, after the treatment

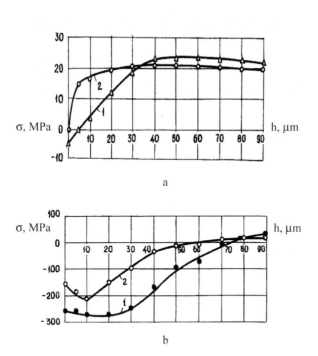

**Fig. 9.3** Distribution of residual stresses with depth for treated blades of the titanium-based VT 8 alloy. a, the ion-plasma nitriding of the surface: 1, the input edge of the blade, 2, the output edge of the blade. b, as a and with the treatment of the surface by ball-bearings in the ultrasonic field: 1, the input edge of the blade, 2, the output edge.

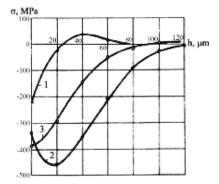

**Fig. 9.4** Distribution of residual stresses with depth for compressor blades:1, vibratory polishing; 2, as 1 and with treatment by ball-bearing in the ultrasonic field; 3, as 1 and shot peening.

by ball-bearings in the ultrasonic field the stresses become negative (−150 to −250 MPa) and are distributed down to a depth of 90 μm.

Similar data from [58] are presented in Figure 9.4. We draw attention to the curve 1 in Figure 9.4 for the vibratory polished specimen. There is an area with tensile residual stresses at a depth of 40 μm. This area can initiate a fatigue crack.

Residual stresses on the surface of a blade are affected by the employment of technology. The influence of the time of the shot peening process on residual stresses is illustrated in Figure 9.5. The fine input edges of the blades become more stressed than the output ones.

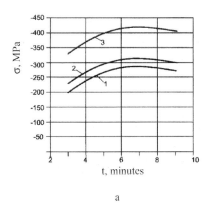
a

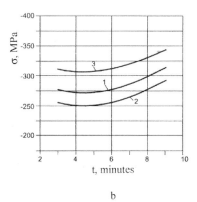
b

**Fig. 9.5** The effect of time of shot peening on the surface residual stresses for compressor blades: a, input edges of blades; b, output edges of blades. 1, the treatment under a pressure of 0.08 MPa; 2, 0.10 MPa; 3, 0.12 MPa.

The surface structures which have been created by shot peening or by ball-bearings treatment in the ultrasonic field can be expected to extend the fatigue lifetime.

Figure 9.6 presents an example of the field of the compressive residual stresses near the surface of structural steel.

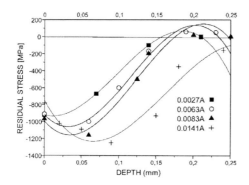

**Fig. 9.6** The field of the compressive residual stresses after shot peening: ■, intensity of shot peening equals 8 psi; ○, 13 psi; ▲, 18 psi; + , 45 psi. Steel AISI 4340 (0.4C+1.74Ni+0.8Cr+0.25Mo+0.25Si,wt%). Reprinted from [59] with permission from Elsevier.

An increase in the shot peening intensity results in an increase in the maximum compressive residual stress up to 1200 MPa. The depth of the favorable stress field extends up to 250 μm.

One can see in three of the curves in Figure 9.6, a maximum in the tensile residual stresses at a depth of 0.17–0.22 mm. These stretched areas that are located at subsurface layers can be the origin of a fatigue crack. The fourth curve has no maximum.

The manufacturing procedure requires X-ray control of the component surface after the engineering processes responsible for introducing the residual stresses and the material behavior.

The physical foundation of the neutron diffraction method is analogous to the X-ray technique, see e.g. [60]. The residual strain is obtained from measurements of a change $\Delta d$ in the interplanar spacings $d$ of the crystal lattice. When a beam of neutrons of fixed wavelength is incident upon a metallic specimen, a diffraction pattern with a sharp peak is produced. The change in the $d$ value due to a residual stress causes a shift in the peak. The neutron diffraction method is especially applicable to components with a complicated shape and to cumbersome details. One applies the neutron diffraction technique to determine the undesirable stresses introduced by welding and the beneficial residual stresses generated by shot peening or other treatment.

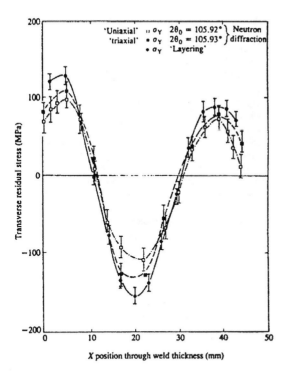

**Fig. 9.7** The transverse residual stress distribution across a multi-pass butt weld in an aluminum-based alloy. Reprinted from [60] with permission from Elsevier.

Welds are known to be a source of detrimental tensile residual stresses. A representative transverse distribution of residual stresses in a joint weld is presented in Figure 9.7. The data were obtained by the neutron diffraction technique. The residual stress changes its sign twice while moving through the weld thickness. Near the weld, the tensile residual stress culminates at 120 MPa. For an aluminum alloy, the value is as large as the yield stress. This is just the area of the heat effect of the joint weld. The same area is frequently a source of the fatigue failure.

The initiation of a crack in subsurface layer of a Ti-based alloy is shown in Figure 9.8. A peak of the tensile residual stress causes the initiation of fatigue crack.

There is a danger of an over-work hardening during processing of the severe plastic strain of the surface. It is sometimes appropriate to apply a softening temper to the material after strengthening.

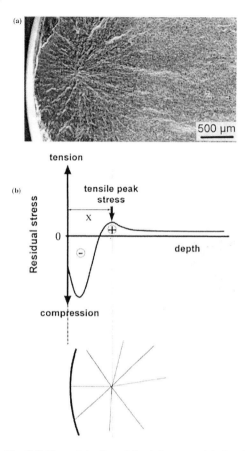

**Fig. 9.8** The origination of the fatigue crack in tensile areas of a fatigued specimen: a, the subsurface crack nuclei in the Ti+6Al+7Nb alloy; the specimen was treated by roller-burnishing; b, schematic representation of the residual stress distribution. Reprinted from [61] with permission from Elsevier.

## 9.2
### Fatigue Strength of the Strained Metallic Surface

There is now a considerable body of evidence that nanostructuring of metallic surfaces enhances their fatigue strength.

The fatigue life of the stainless 316L steel was found to increase considerably compared to the untreated material [62]. The fatigue test data are shown in Figure 9.9.

One can see that, after nanostructuring, the $S - N$ curves move upwards in comparison with the curve for untreated specimens. The nanostructured material reveals an improvement in the fatigue limit of 21%. The treatment of surface by shot balls of 3 mm in diameter is more beneficial than treatment by 2 mm balls. The nanostructured and annealed material displays the bet-

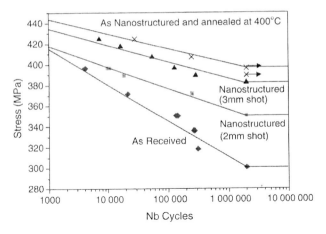

**Fig. 9.9** The stress amplitude versus the number of cycles to fracture (S – N diagram) for the 312L stainless steel. The cycling frequency is 10 Hz. Reprinted from [62] with permission from Elsevier.

ter properties than the nanostructured only. The ratio of lifetimes between nanostructured and annealed steel and nanostructured only is equal to 3.5 or 7.1 under strain amplitudes of 430 and 390 MPa, respectively. Relaxation of the over-stressed areas is likely to take place at 673 K (400°C). The compressive residual stresses remain, as yet, at this temperature.

The authors of [63] have studied the influence of deep rolling on the properties and fatigue behavior of the Ti+6Al+4V alloy. Deep rolling involves the plastic deformation of the surface layers of a specimen by a spherical or cylindrical rolling element under controlled pressure.

It is thought that the deep rolling process has an advantage over shot peening because this process is relatively inexpensive and allows one to achieve a smoother surface topography, whereas the compressive residual stresses retard fatigue-crack initiation and propagation. The authors obtained the surface roughness parameter $Rz = 0.8$ μm for the deep-rolled titanium-based alloy. (Our result after shot peening is equal to 1.75 μm (Figure 9.1c). However, our data have been obtained for the surfaces of real compressor blades, whereas the authors [63] handled cylindrical specimens with a gauge length of 15 mm and a diameter of 7 mm).

Fatigue tests were performed at a frequency of 5 Hz at temperatures of 298 K (25°C) and 723 K (450°C). A comparison of the fatigue behavior of the virgin and the deep-rolled alloy is presented in Figure 9.10. Deep rolling results in a significant enhancement in the fatigue lifetime at room temperature. Under a load of 500 MPa the fatigue life increases by two orders of magnitude. The increase in lifetime at 723 K is considerably lower.

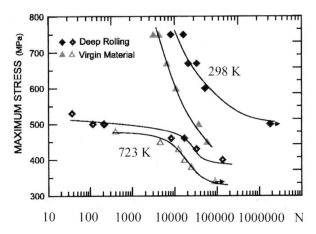

**Fig. 9.10** Dependence of the maximum stress on the number of cycles to fracture $N$ for the Ti+6Al+4V alloy. The beneficial effect of deep rolling at two temperatures can be easily seen. Reprinted from [63] with permission from Elsevier.

Scanning electron microscopy evidence confirms that deep rolling is effective in retarding the initiation and initial propagation of fatigue cracks. Deep rolling has a positive influence on fatigue properties by lowering the growth rate of the initial fatigue crack, by a factor of 2–3 compared with corresponding behavior in the virgin material. The authors of [63] believe that the greater effectiveness of deep rolling is attributed to a higher magnitude of the induced compressive stresses, a higher degree of work hardening, and a significant decrease in surface roughness.

However, it is clear that for components of a complex shape such as gas-turbine blades and disks the shot pining process is more applicable than the deep rolling.

Processes of shot peening and roller burnishing lead to a considerable improvement in fatigue performance for a Ti-based alloy (Figure 9.11). The increase in the fatigue strength for an alloy with larger yield strength (Figure 9.11b) is much higher than for an alloy with lower yield strength (Figure 9.11a).

Fatigue lifetime depends on processing as well as on test conditions. This statement is illustrated by Figure 9.12. Some authors [59] evaluated the fatigue strength of AISI 4340 steel as a function of the conditions of shot peening. Four operating modes of the treatment were applied. These modes were different in intensity of shot peening (see Figure 9.6). There is no difference between the treatment results under a relatively high stress level of 1360 MPa. However, from Figure 9.12 the advantage of the intermediate operating mode over the severe one under test stresses of 830–930 MPa, is obvious.

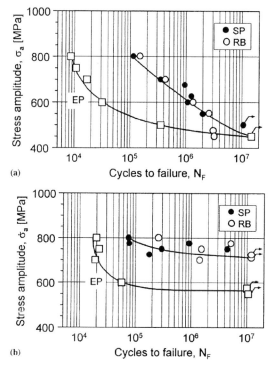

**Fig. 9.11** Effect of surface treatments on S–N curves for the Ti+6Al+7Nb alloy: EP, electrolytically polished surface; SP, shot peening; RB, roller burshing; (a) $\sigma_{0.2}$ of the alloy equals to 920 MPa; (b) $\sigma_{0.2}$ of the alloy equals 1030 MPa. Test frequency is about 50 Hz. Reprinted from [61] with permission from Elsevier.

The austenitic iron–nickel alloy Invar contains 36% Ni. A unique property of this alloy is its low coefficient of thermal expansion. However, the alloy has a relatively low strength and fatigue life. The authors of [64] have employed the severe plastic strain in order to enhance the strength parameters without sacrificing the low thermal expansion coefficient. The authors have used the equal-channel angular technique to impose extremely large stresses on bulk samples. Billets of square 14 mm × 14 mm cross-section have been subjected to multiple pressing in an installation having two square channels intersecting at 90°.

This treatment results in a significant decrease in the grain size. Figure 9.13 illustrates the effect of the severe plastic strain on the structure of the alloy. The Figure clearly shows the formation of the fine and homogeneous structure. The authors estimate the mean grain size as 300, 260 and 180 nm for a, b, c, of Figure 9.13, respectively. The fatigue lifetime after the treatment is found to be by an order of magnitude longer (Figure 9.14).

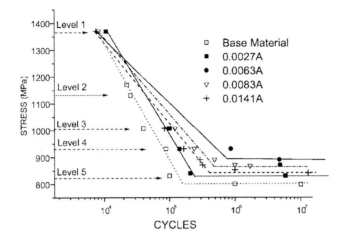

**Fig. 9.12** S–N comparative curves of the base materials (□) and after shot peening treatment: ■, intensity of shot peening is 8 psi; ○, 13 psi; ▲, 18 psi; +, 45 psi. The AISI 4340 steel is under investigation. Reprinted from [59] with permission from Elsevier.

## 9.3
## Relaxation of the Residual Stresses under Cyclic Loading

The improvement in the fatigue lifetime is known to depend on the stability of the induced residual stresses and hardening in the near-surface region. The beneficial residual stresses are affected by cyclic loading. There are experimental data confirming the fact that the stresses can redistribute and relax due to repeated alternating loading. At the same time, it is important to understand how the compressive residual stresses are affected by real-time operation.

We have studied the distribution of compressive stresses near the surface of the compressor blades before and after operation in a gas-turbine engine

**Table 9.2** Distribution of residual stresses at the surface of compressor blades and redistribution during operation. $\sigma_{sur}$ is the residual stress at the surface; $\sigma_{min}$ is the minimum value of the residual stresses; $h$ is the depth of the stress distribution; USF is the ultrasonic field.

| No Processing | Input edges | | | Output edges | | |
|---|---|---|---|---|---|---|
| | $\sigma_{sur}$ (MPa) | $\sigma_{min}$ (MPa) | h (μm) | $\sigma_{sur}$ (MPa) | $\sigma_{min}$ (MPa) | h (μm) |
| 1 Vibration polishing + shot peening | −450 | −450 | 100 | −320 | −320 | 100 |
| 2 Nitriding + treatment by bearing balls in USF 300 s | −260 | −280 | 80 | −150 | −210 | 60 |
| 3 As 2 + operation of 980 hours | −200 | −140 | 70 | −150 | −250 | 70 |
| 4 Vibration polishing + treatment by ball-bearings in USF + operation for 1500 hours | −350 | −350 | 70 | −410 | −410 | 70 |

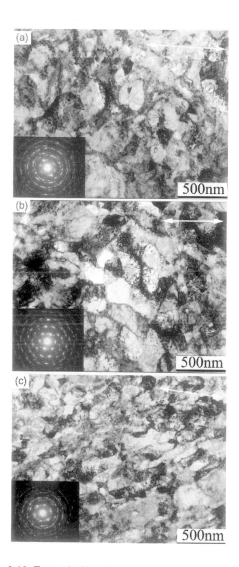

**Fig. 9.13** Transmission electron micrographs of the Invar alloy after equal-channel angular pressing: a, two passes; b, 8 passes; c, 12 passes. Reprinted from [64] with permission from Elsevier.

for 980 and 1500 hours. It is impossible to measure residual stresses using the same blade before and after its operation in a gas-turbine engine. We have therefore studied different blades in various states before and after running.

Figures 9.15, 9.16, and Table 9.2 illustrate the data on redistribution of the residual stresses as a result of the operation of the blades.

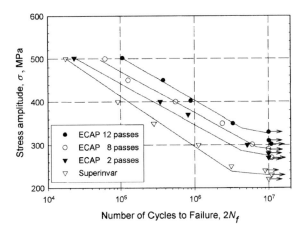

**Fig. 9.14** The effect of stress amplitude on the number of cycles to fracture for specimens after equal-channel angular pressing (ECAP) for the Invar alloy, the frequency of tests is 10 Hz. Reprinted from [64] with permission from Elsevier.

The considerable relaxation of induced compressive stresses during operation is unlikely to occur in practice. The operation leads only to a decrease in the stress depth of from 80–100 to 70 µm. A decrease in the absolute stress value is more for the input edge of the blades than for the output one. There is no noticeable difference between operations in the gas-turbine during 980 or 1500 hours.

A characteristic feature of the curves $\sigma - h$ is a minimum. In our case the surface nitriding or polishing and subsequent strengthening by ball-bearings in the ultrasonic field for 600 s turns out to be the best technology (Figures 9.15 and 9.16, curves 3 and 4).

The authors of [65] carried out a series of three experimental fatigue tests on standardized specimens: torsion, rotary bending and tension – compression. The examination was conducted with two low-alloy steels. Cylindrical specimens were first shot-peened, and then fatigue-tested. The thickness of the stressed peened layer was equal to 300 µm.

Figure 9.17 shows the compressive residual stress distribution before and after fatigue tests. As in our experiments (Figures 9.2b, 9.3b, 9.15) the minimum of the initial compressive stresses is observed below the surface. The initial residual stresses change with stress amplitude during the fatigue test. The graphs indicate that the residual stresses are released and redistributed during cycling. The relaxation depends on the amplitude and direction of the

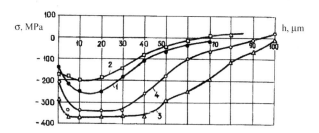

**Fig. 9.15** Redistribution of residual stresses at the surface of strengthened blades after the operation for 980 hours in a gas-turbine engine. The previous treatments are: 1 and 2, the ion-plasma nitriding and strengthening by ball-bearings in the ultrasonic field for 300 s; 3 and 4, the ion-plasma nitriding and strengthening in the ultrasonic field for 600 s; 2 and 4, input edges of blades, 1 and 3, output edges.

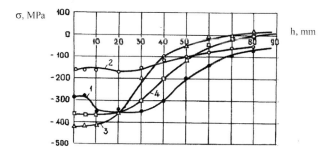

**Fig. 9.16** As in Figure 9.15 after the operation for 1500 hours in a gas-turbine engine. The previous treatments are: 1 and 2, glossing and strengthening by ball-bearing in the ultrasonic field for 600 s; 3 and 4, vibration polishing and strengthening in ultrasonic field, 600 s; 2 and 4, input edges of blades, 1 and 3, output edges.

applied load. The curves shift by a value of $+100$ to $+250$ MPa after $10^3$ cycles. The higher the residual stress module the greater the relaxation. However, it is clear that compressive stresses for the most part, remain after cycling. A level of residual stresses of $-400$ to $-500$ MPa is observed also for cycled specimens.

It is a of common observation that the area of thermal influence near a joint weld is in danger of failure. The evolution of residual stresses near a weld as a result of cycling is shown in Figure 9.18. In this case a high heat input of 56 kJ mm$^{-1}$ leads to lower residual stresses because of the annealing effect. The area with positive stresses is situated in an as-welded specimen at a distance of 6–18 mm from the weld center.

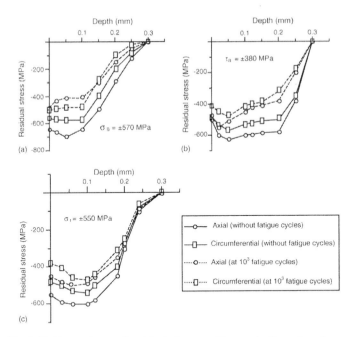

**Fig. 9.17** Dependence of residual stress profiles on depth and mode of a subsequent cycling for the 0.35C+4.5Ni+1.8Cr+0.5Mo steel. The fatigue tests are: a, rotary bending; b, torsion; c, tension-compression. Reprinted from [65] with permission from Elsevier.

100 cycles of fatigue loading were then applied to the welds by bending with peak stresses of 100, 150 or 200 MPa. The increment in the positive residual stresses as a consequence of cycling is +40 to +50 MPa at a distance of 8 mm from the weld. The weld toe, which is the usual crack initiation site in fusion welds, lies about the same distance from the weld center line.

The temperature of the cycling is found to have a noticeable influence on the residual stress relaxation. Simple thermal exposure (with no cycling) of a deep-rolled titanium-based alloy causes a substantial reduction in the near-surface residual stresses [63].

The redistribution of compressive residual stresses during cycling of the alloy at different temperatures is shown in Figure 9.19. The authors consider the half-width of X-ray reflections that remain stable to thermal and mechanical excursions (Figure 9.19b). The compressive stresses at the 500 μm depth progressively degrade with fatigue cycling temperature. The maximum residual stress, which is at a depth of 70 μm, decreases after 50% of the lifetime at 450°C from $-950$ to $-200$ MPa.

The authors conclude that the benefit of a high-temperature fatigue strength is primarily associated with the formation of a nano-scale microstructure as

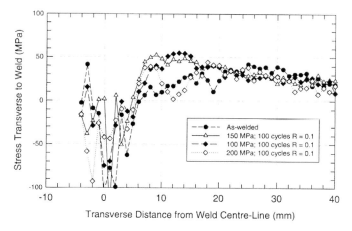

**Fig. 9.18** Residual stresses versus transverse distance for the welded joint of an aluminum alloy. Reprinted from [66] with permission from Elsevier.

consequence of deep rolling. This microstructure is stable during cycling up to 450°C.

We can also add a possible beneficial influence of microscopic stresses, which are known to affect the X-ray reflection width as much as fine grains.

The relaxation of residual stresses during fatigue cycling of the AISI 4340 steel[1] is shown in Figure 9.20. The progressive relaxation is noted with $10^4$ and $10^5$ cycles. In all cases the compressive stresses do not disappear completely.

According to [67] the macroscopic stresses are generally more stable than the microscopic stresses under identical loading conditions for AISI 1080 steel[2]. The authors note that in high-stacking fault energy materials (e.g. ferrite) with a strong propensity for cross-slip, the dislocation cell structure is formed during cycling. This process is independent of prior cold work hardening. The greater stability of macroscopic stresses is attributed to their longer range nature, which decreases the likelihood of dislocation – dislocation interaction.

In this work, the geometry for the initiation specimens was chosen to simulate fatigue loading in the presence of notch stress concentrators. Holes of 6.08 mm diameter were drilled in a 6 mm thick bar. After polishing and annealing the holes were radially expanded by hydraulically pressing a 6.35 mm WC ball. This press fitting operation was found to lead to compressive macroscopic stresses of −750 MPa. They remained relatively stable throughout the

---

**1)** Composition of AISI 4340 steel used was (wt.%)
0.41C+0.73Mn+0.8Cr+0.25Mo+0.25Si.

**2)** Nominal composition of AISI 1080 steel is (wt%) 0.8C + 0.75Mn + max0.04P + max0.055S.

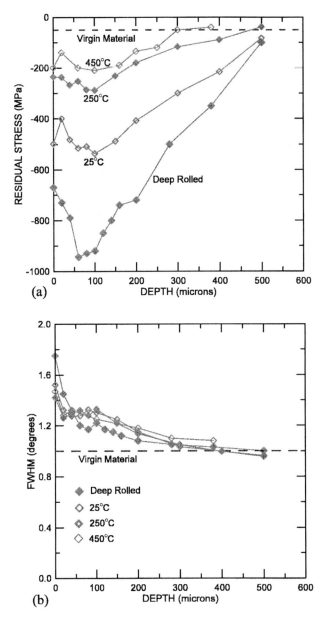

**Fig. 9.19** Redistribution of residual stresses (a) and the half-width of the X-ray reflection (b) for the Ti-based alloy before and after fatigue cycling at 25, 250, and 450°C. The frequency of test 5 Hz. Reprinted from [63] with permission from Elsevier.

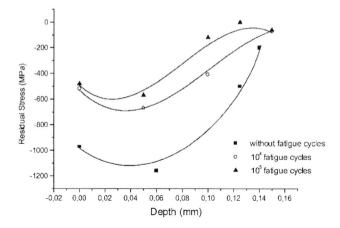

**Fig. 9.20** Relaxation of residual stresses in structural steel as a result of cycling. The test frequency is 50 Hz. Reprinted from [59] with permission from Elsevier.

fatigue life. After $3 \times 10^3$ cycles the stresses were $-600$ MPa and thus fatigue-crack initiation was prevented.

## 9.4 Microstructure and Microstructural Stability

It is of interest to consider evolutions of the structure during fatigue tests. It can be seen from Figure 9.21a that the initial microstructure of the EP479 superalloy (Table 2.1) is martensite that consists of parallel plates. The microstructure contains also austenite and carbide of type $Cr_{23}C_6$. Dislocations are generated as a result of the alternating mechanical stresses. The martensite begins to decay gradually and subgrains begin to form as can be seen in Figure 9.21b and c. Further, an equiaxial structure is formed. The size of subgrains varies from 100 to 200 nm, Figure 9.21d.

The mechanical and thermal stability of nanocrystalline layers has attracted the attention of many investigators.

The authors of [62, 68] studied the effect of a surface mechanical attrition treatment on the structure and properties of austenitic 316L steel. This well known stainless steel contains (in mass %) 0.002 C, 17.1 Cr, 12.0 Ni, 2.0 Mo, 1.7 Mn. The steel in its as-received state had an initial grain size of between 10 and 50 µm. To achieve surface nanocrystallization, mechanical attrition treatment by steel shot (3 mm in diameter) in an ultrasonic field was applied. The vibration frequency was 20 kHz and the amplitude of the ultrasonic source oscillations was 25 µm. The treatment time ranged from 5 to 30 min.

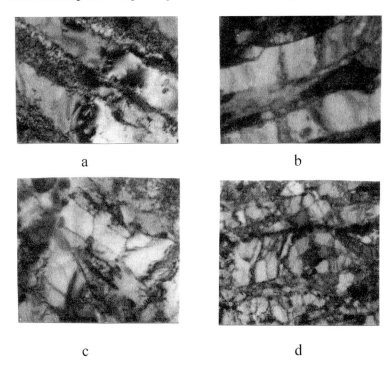

**Fig. 9.21** Transmission electron micrographs showing a change in the microstructure of the EP479 superalloy during fatigue tests. The stress amplitude equals 441 MPa, the frequency is 381 Hz. The number of cycles until fracture equals $2 \times 10^6$; a, initial state, plates of martensite; b, c, decay of martensitic plates, increase in the dislocation density; d, interaction of dislocations with each other and formation of sub-boundaries and subgrains. ×25 000.

A transmission electron microscopic investigation, after the treatment, reveals a nanostructured surface layer of 40 μm deep with a grain size of about 20 nm. The microstructure is characterized by uniformly distributed grains. Figure 9.22 presents the transmission electron microscope pattern of the steel subjected to the surface attrition treatment. Randomly oriented grains of nanometeric dimensions are the reason for the full rings instead of spots in the electron-diffraction patterns.

At the same time, a phase transformation from austenite to martensite, $\gamma \rightarrow \alpha$, occurs partly under plastic deformation. The volume fraction of martensite was determined by the X-ray method after nanostructuring to be about 15%. Moreover, as the stainless steel is a material of low stacking fault energy, the plastic deformation also results in the mechanical twinning. At 50 μm below the surface, twin-twin intersections occur to sustain the higher strain induced by peening (Figure 9.23).

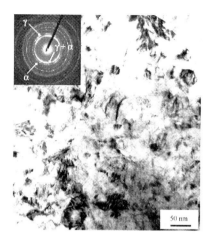

**Fig. 9.22** Transmission electron micrograph showing the nanostructure of the treated surface of stainless 316L steel. Inset is the electron-diffraction pattern, indicating reflections of austenite and martensite. Reprinted from [68] with permission from Elsevier.

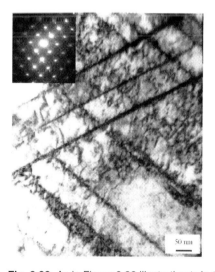

**Fig. 9.23** As in Figure 9.22 illustrating twin-twin intersections.

The authors found at about 200 μm depth from the surface, the presence of unidirectional parallel mechanical twins that lead to the formation of a twin-matrix alternative lamellar structure. A high density of dislocations can also be observed inside the twins and close to their boundaries. Arrangement of these dislocations leads to the formation of walls inside the microscopic twins, while others are arranged into planar arrays.

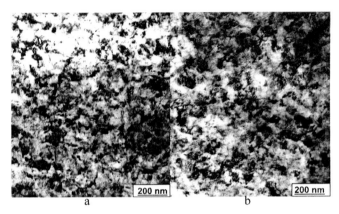

**Fig. 9.24** Transmission electron micrographs of the nanostructured layer of AISI 304 stainless steel: a, after 1 cycle; b, after 8000 cycles (0.93 of the lifetime). $\sigma_a$=350 MPa, T=298 K. Reprinted from [69] with permission from Elsevier.

The authors studied the thermal stability of the nanometer scaled microstructure in the temperature range from 373 to 1073 K (100–800°C). The nanocrystalline layer was found to be stable up to 600°C. The average grain size rose from 10 up to 25 nm during annealing for 10 min at 700°C. The strength properties of stainless steel did not decrease up to 600°C. Annealing between 300 and 500°C improved both the ductility and strength.

The authors of [69] have studied the stability of the near-surface microstructure of the deep-rolled austenitic steel AISI 304 and the Ti+6Al+4V alloy. After deep rolling the steel the grain size in the nanocrystalline layer was about 30 nm directly at the surface and then increased continuously to 40 nm at a depth of 1.4 µm. The subsurface exhibits a high dislocation density in the austenitic matrix between martensitic needles and twins.

The isothermal fatigue experiments were performed using a frequency of 5 Hz at temperatures from 298 to 873 K (25 – 600°C). Figure 9.24 demonstrates that the nanocrystalline layer remains stable during low-cycle tests at room temperature. It is evident that cyclic loading did not affect the nanocrystalline layer significantly. However, there is transmission electron microscope evidence that the subsurface non-nanocrystalline layer with high initial dislocation density is strongly affected by fatigue. The dislocation density decreases with increasing number of cycles. X-ray study indicates the softening of steel (Figure 9.25).

The stability of the nanocrystalline surface layer in a titanium-based alloy was determined in fatigue tests conducted at different temperatures and a stress amplitude of 430 MPa. The grain size within the whole nanocrystalline layer had also not changed significantly after fatigue at 723 K. However, the

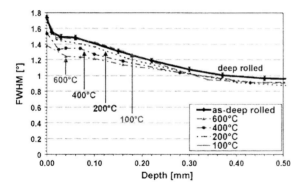

**Fig. 9.25** Dependence of the full width at half maximum (FWHM) of an X-ray reflection on the test temperature for stainless steel. $\sigma_a = 280$ MPa, 2000 cycles. Reprinted from [69] with permission from Elsevier.

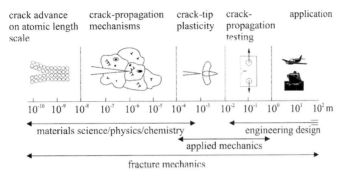

**Fig. 9.26** Levels of approach to the fatigue failure problem. Scientific disciplines are indicated. Reprinted from [70] with permission from Wiley-VCH.

authors note that the results obtained from their investigation may not be valid for other test frequencies. Indeed, 2000 cycles at 873 K (the oscillation frequency is 5 Hz) correspond to an exposure interval of only 6.67 minutes. This time seems to be insufficient for diffusion processes of softening. In high-cycle fatigue a material undergoes millions of cycles.

## 9.5
## Empirical and Semi-Empirical Models of Fatigue Behavior

It is a matter of general observation that a fatigue phenomenon consists of two stages: the initiation of a crack and the crack growth. Both processes are significant.

# 9 Nanostructuring and Strengthening of Metallic Surfaces. Fatigue Behavior

The prediction of fatigue properties is a complex problem on account of large quantity of factors that have an influence on the process.

Various models were developed for the prediction of the rate of the crack propagation. The physical understanding of the fatigue phenomena is essential for the fatigue prediction.

From the physical point of view there are three levels of approach to the fatigue problem: macroscopic, microscopic, and atomic. From the first one the material is considered to be a homogeneous continuum. The second approach operates with microscopic concepts: microstructure, microscopic crack, fracture. The atomic consideration of the problem deals with a dislocation density evolution, with a part of vacancies, generation of surface steps, distribution of microscopic stresses, variation in the grain and subgrain structures. For the third approach, on the atomic scale, there is not enough experimental information.

These approaches do not exclude each other. Some models have characteristic features of several approaches.

Figure 9.26 illustrates visually that the fatigue failure problem is investigated in different ways depending on the length scale, for which structure elements are considered.

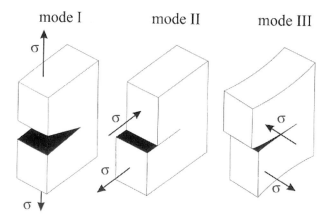

**Fig. 9.27** Scheme of the three strain modes.

## 9.5.1
### Fatigue-Crack Propagation in Linear Elastic Fracture Mechanics

An approach of the elastic fracture mechanics does not take into consideration the atomistic structure of the fatigued material.

The discussion of the fatigue-crack propagation is based on the theory of linear elastic and nonlinear fracture mechanics in a uniform medium. The appropriate conditions for the dominance of critical fracture parameters are obtained from a knowledge of the accuracy of asymptotic continuum solutions [71]. This approach is successful due to the engineering design.

The main concept of elastic fracture mechanics is the stress intensity factor $K$. It is a measure of the intensity of the field near the crack tip under linear elastic conditions. The initiation of crack advance under a monotonic loading condition is characterized by the critical value of the stress intensity factor $K_c$. The value of $K_c$ is a function of the material microstructure, the temperature, the mode of loading, and the strain rate.

There are different modes of fracture (Figure 9.27). Mode I is the tensile opening mode, in which the crack faces are separated from each other in a direction normal to the plane of the crack. The mode of loading is called the tensile mode if the crack surfaces move directly apart. The stress intensity factor is commonly expressed in terms of applied stresses. In Figure 9.28 the values of $K_I$ are shown for the tensile mode of loading in a material that contains a crack.

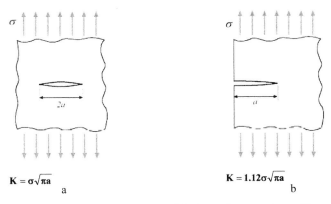

**Fig. 9.28** Examples of the calculation of the intensity stress factor $K_I$: a, the crack is situated in the bulk of a material; b, the crack is located near the surface.

Mode II is the in-plane sliding mode, in which the crack faces are mutually sheared in a direction normal to the crack front. Mode III is the tearing or anti-plane shear mode in which the crack faces are sheared parallel to the crack front. The crack face displacements in modes II and III are analogous to the motion of edge and screw dislocations, respectively.

Every mode of the fracture has its own intensity stress factor denoted by $K_I, K_{II}, K_{III}$, respectively.

Irwin [72] introduced the energy release rate $G_I$. By definition

$$G_I = -\frac{dW}{dA} \tag{9.1}$$

where $dW$ is a change in the total energy of the cracked body, $dA$ is the increment in the crack surface. It turned out that the energy release rate $G_I$ is related to the stress intensity factors in the following way

$$G_I = \frac{1-v^2}{E}(K_I^2 + K_{II}^2) + \frac{1+v}{E}K_{III}^2 \tag{9.2}$$

where $E$ is Young's modulus and $v$ is the Poisson coefficient.

The authors of [74] assumed that the growth rate of a crack under cyclic loading obeys the law

$$\frac{da}{dN} = C\,(\Delta K)^m \tag{9.3}$$

where $a$ is the length of the fatigue crack, $N$ is the number of cycles, $\Delta K$ is the stress intensity factor range and $C$ and $m$ are empirical constants. The values of $C$ and $m$ are dependent on the material properties and microstructure, the mean load ratio, the loading frequency, the loading mode, the test temperature, and also on the environment influence.

Equation (9.3), the so-called Paris formula, is a plausible assumption for the measurement results of the propagation rate of a fatigue crack.

$\Delta K$ is defined as

$$\Delta K = K_{max} - K_{min} \tag{9.4}$$

where $K_{max}$ and $K_{min}$ are the maximum and minimum stress intensity factors, respectively. The values of $K_{max}$ and $K_{min}$ correspond to the maximum stress, $\sigma_{max}$, and the minimum stress, $\sigma_{min}$.

For a crack located in the center of a plate (see Figure 9.28a) $K_{max} = Y\sigma_{max}(\pi a)^{\frac{1}{2}}$ and $K_{min} = Y\sigma_{min}(\pi a)^{\frac{1}{2}}$, where $Y$ is a size-correction coefficient for the plate and $2a$ is the crack length. The value of the stress intensity factor, below which the inadmissible growth of the crack does not occur, is called the threshold stress intensity factor $K_{th}$.

Figure 9.29 illustrates a plot of $\ln(da/dN)$ versus $\ln \Delta K$. One identifies three distinct regimes of the crack growth in this graph. In regime A the average growth increment per cycle is smaller than an interatomic spacing in the crystal lattice. If the stress intensity factor is less than the threshold $\Delta K_{th}$ the crack remains latent. Regime B exhibits a linear dependence of $\ln(da/dN)$ on $\ln\Delta K$, (9.3). Regime C is related to the range of high values of $\Delta K$ where rates of the crack growth increase rapidly, causing a catastrophic failure.

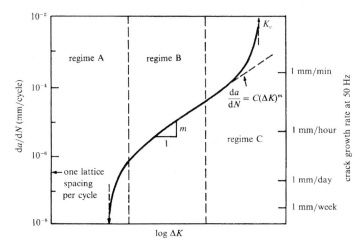

**Fig. 9.29** Schematic illustration of the different regimes of fatigue-crack propagation.

In practice, $K_{th}$ is usually assumed as the value of the stress intensity factor, at which the crack growth rate is of the order of 0.1 nm cycle$^{-1}$ [74]. The typical range of the threshold intensity factor for metals is from 3 to 300 MPa m$^{\frac{1}{2}}$. $K_{th}$ values were reported [95] to be equal to 7.0 and 2.9 MPa m$^{\frac{1}{2}}$ for aluminum-based and titanium-based alloys, respectively.

Therefore the Paris formula (9.3) is an empirical relationship between the crack growth rate and the loading. The values of C and m do not have any clear physical meaning; they are fitting parameters to be determined experimentally from fatigue tests.

Krupp [70] believes we can distinguish between short and long cracks. Moreover, the author believes that we should define three types of short crack. The propagation rate $da/dN$ immediately after growth commences is determined by interactions with the local microstructural features, which are characteristics of the material, e.g. grain boundaries and precipitates. Such cracks are termed microstructural short cracks. They are strongly affected by the microstructure. The influence of microstructure vanishes once the crack length exceeds several grain diameters. Crack propagation is driven by the plastic zone ahead of the crack tip. These cracks are termed mechanically short cracks. Cracks termed physically short are characterized by a negligibly small plastic zone as compared with the crack length. Concepts of linear elastic fracture mechanics are applicable to physically short cracks. A fatigue crack is considered to be a long crack when, besides the intrinsic crack-driving, the extrinsic influence factors are fully developed. The author assumes that the length of the long cracks is more than 0.5 mm.

The Paris equation (9.3) does not hold for short fatigue cracks. In steels, aluminum-based, and nickel-based alloys an initially decreasing crack growth rate has been observed. Concepts of linear elastic fracture mechanics are not applicable for the length scale of short cracks because the material cannot be considered to be a homogeneous continuum.

An example of the crack growth rate dependence on $\Delta K$ for aluminum-based alloys is shown in Figure 9.30. The Paris equation is obeyed for specimens with a polished surface as one can see in Figure 9.30a.

However, this equation fails for specimens with a shot-peened surface, Figure 9.30b. In the last case there is no any 'well behaved' connection between

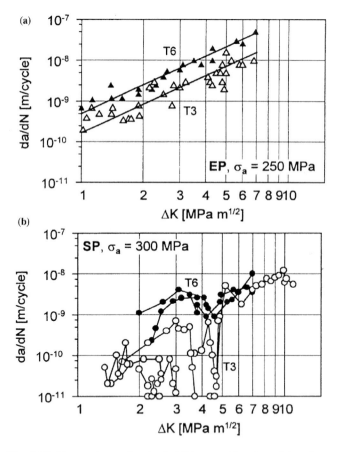

**Fig. 9.30** The crack growth rate $da/dN$ versus the stress intensity factor $\Delta K$ for aluminum alloys. T3 is the Al-based alloy with $\sigma_{0.2} = 360$ MPa; T6 is the Al-based alloy with $\sigma_{0.2} = 240$ MPa; a, the surface of the specimens was electrolytically polished (EP); b, the surface of the specimens was subjected by shot peening (SP). Reprinted from [61] with permission from Elsevier.

two values. The experimental points of the dependence $\ln(da/dN) - \ln(\Delta K)$ are situated chaotically.

Some researchers have attempted to introduce structure factors into their models. The authors of [75] made the assumption that, during the fatigue loading, the material ahead of the crack tip undergoes plastic deformation. Tensile loading during the loading part of the cycle and compressive loading during the unloading part of the cycle will cause either cyclic hardening or softening of the material. As a result of this cyclic hardening or softening, the local cyclic yield strength of the material at the crack tip region $\sigma'_{yz}$ will be different from its monotonic yield strength. The authors assume that a source of dislocations S exists at the center of a grain. During cyclic loading, the source S continues to emit dislocations. These dislocations move in intersecting planes and are blocked by the grain boundary. At the threshold level, the dislocations are unable to break through the grain boundary. Authors further assumed that the slip band length equals half of the grain diameter. $m$ in the Paris formula can be approximated by $\sqrt{3}$. Finally they obtain an expression for $K_{th}$ in the form

$$\Delta K_{th} = 84\sigma'_{yz}(1-R)\sqrt{D} \tag{9.5}$$

where $R$ is the load ratio[3)].

It follows from (9.5) that the fatigue threshold increases with increasing grain size of the material as $\sqrt{D}$. The authors did not reported whether they obtained the confirming experimental data. The statement that the smaller the grain size the smaller the threshold stress intensity factor, contradicts the generally accepted opinion of the endurance strength of nanostructures.

The author of [77] assumed that the threshold condition is determined by a slip band arising just ahead of the crack tip. He derived the formula

$$\Delta K_{th} = Y(2\pi r)^{\frac{1}{2}} \tag{9.6}$$

where $Y$ is the flow stress of the material and $r$ is the length of the slip band.

### 9.5.2
### Crack Propagation in a Model Crystal

An atomic model of a crack propagation is considered in [76].

The energy release rate $G$ (J m$^{-2}$) is defined as the elastic energy released per unit area in advance of the crack tip,

$$G = \frac{1}{2}\frac{EW}{1-v^2}\varepsilon^2, \tag{9.7}$$

---

3) The load ratio $R = \sigma_{min}/\sigma_{max}$. With this definition, $R = -1$ for the fully reversed sinusoidal loading, $R = 0$ for the zero-tension fatigue, and $R = 1$ for the static load.

where $E$ is Young's modulus, $\nu$ is the Poisson ratio, $\varepsilon$ is the strain, and $W$ is the specimen size along [001]. The crack propagates when $G$ becomes equal to or greater than $2\gamma$, where $\gamma$ is the surface energy of each plane of the crack.

According to the molecular dynamics method, one specifies the positions and velocities of all atoms at zero time. The Newton equations of motion are integrated using a numerical algorithm. Newton's second law is given by

$$F_i = m_i \frac{dv_i}{dt}; \quad F_i = -\frac{\partial U_{pot}}{\partial x_i}, \qquad (9.8)$$

where $m_i$, $x_i$, $v_i$ are the mass, the coordinate, and the velocity of an $i$ atom, $U_{pot}$ is the total potential energy of the system.

Crack propagation was studied by the authors on a (001) plane of the nickel single crystal (Figure 9.31). Authors examined the crack system (001) [100] in a slab that contained $1.6 \times 10^5$ atoms. The crystal was first strained by a load that corresponded to the Griffiths value $G_0 = 2\gamma = 3{,}77\, \mathrm{J\,m^{-2}}$. Each current value of the load[4] $G_s$ depending on the strain $\varepsilon$ was calculated from (9.7). In the case under study, $W = 17.4\,\mathrm{nm}$ and $G_s$ only depends on the strain, $G_s = 2949.4\,\varepsilon^2\,\mathrm{J\,m^{-2}}$. In fact, a brittle failure is modeled in this work since a fixed value of temperature (T=10 K) is maintained.

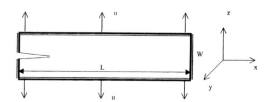

**Fig. 9.31** Scheme of crack propagation under tensile strain loading (mode I) for a model single crystal. Reprinted from [76] with permission.

The velocities of cracks of 792, 1029, and 1482 $\mathrm{m\,s^{-1}}$ were obtained for a load of $1.2G_0$, $1.5G_0$, and $2.0G_0$, respectively. The smallest load which caused a crack to propagate in the defect-free crystal was found to be equal to $G_c = 1.2G_0 = 4.54\,\mathrm{J\,m^{-2}}$.

Two lines of vacancies along [010] were introduced into the model crystal of 160 000 atoms in various positions near the crack tip, as shown in Figure 9.32, c, d, and e. Critical value $G_c$ decreases from $1.2G_0$ to $1.1G_0$. Thus, for the studied model lines of vacancies were shown to enhance strain fields at the crack tip, resulting in the lower rate of the strain energy release.

---

4) The authors make use of the term load. It is obvious that this is energy per unit area.

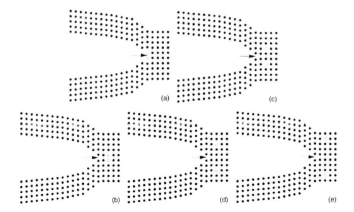

**Fig. 9.32** Sketches of atomic configurations near the crack tip for the model crystal with vacancies. Two lines of vacancies (open circles) are introduced along the y-axis in a symmetric fashion. Reprinted from [76] with permission.

## 9.6 Prediction of Fatigue Strength

Schijve [78] believes that the prediction problems can be divided in two categories. The concept of similarity is characteristic for the first category, and damage accumulation is the base of the second category. This physical principle is the basis of many predictions of material properties. The author assumes that 'it should be realized that this physical principle does not necessarily imply that the physical mechanism of the fatigue phenomenon should be understood'. The phenomenon of fatigue is very complex indeed. It is difficult to compare, meaningfully, data that were obtained by different investigators.

A theoretical model which takes two beneficial effects into account was developed in [65]. These effects are: the compressive residual stresses and strain hardening. The non-cracking condition in the framework of the proposed model is expressed as

$$\xi_a + \alpha P_{max} \leq \beta \tag{9.9}$$

where $\xi_a$ is the maximum amplitude of the second invariant of the stress deviator $[\sqrt{J_2(t)}]$, $\alpha$ and $\beta$ are material parameters and $P_{max}$ is the maximum hydrostatic pressure within a loading cycle. The hydrostatic pressure is equal to zero for a torsion test.

The authors introduce a strain-hardening factor, $C_w$, which is defined as

$$C_w = \frac{b}{b_0} \tag{9.10}$$

where $b$ and $b_0$ are X-ray diffraction widths of the treated and base materials, respectively.

The material parameter $\beta$ is given by

$$\beta = \tau_d \left(\frac{b}{b_0}\right)^{\frac{1}{2}} \tag{9.11}$$

where $\tau_d$ is the fatigue limit in a fully reversed torsion test for the base material.

The second parameter $\alpha$ can be obtained from both fully reversed bending and torsion tests:

$$\alpha = \frac{\tau_d - (\sigma_d/\sqrt{3})}{\sigma_d/3} \tag{9.12}$$

where $\sigma_d$ is the fatigue limit in fully reversed bending test.

In order to use the criterion of (9.9), it is necessary to determine the fatigue limits of the base material in bending as well as torsion tests, and to know the residual stress distribution and width of X-ray reflections.

The low-alloy steel contained (wt. %) 0.37C+4.5Ni+1.8Cr+0.5Mo was treated by shot peening. The values of $\tau_d$ and $\sigma_d$ were found to be equal to 320 and 555 MPa, respectively. The width of X-ray reflections decreases continuously from 3.65° at the surface to 2.2° at a distance 300 μm from the surface; the factor $C_w$ varies correspondingly from 1.82 to 1.01.

Shot peening for the studied condition increases the fatigue limit by 8–22% when compared with the base material. According to the authors' data the improvement in the fatigue life is essentially due to the increasing effect of strain hardening. The nitriding treatment of two structural steels shows an enhancement in the fatigue limit from 15 to 26%. The improvement in the fatigue limit is also attributable to strain hardening. The predicted fatigue limits are in agreement with experimental data.

The authors believe that the increasing effect of strain hardening on the fatigue strength is more important than the influence of compressive residual stresses. This conclusion does not seem very convincing. According to their data (Figure 9.17) compressive stresses kept up to $-400$ MPa at the surface of steel after fatigue tests. The broadening of X-ray reflections is known to be caused by microscopic stresses as well as by small particles (grains). It would be interesting to check the conclusions experimentally. The model predicts the initiation of the fatigue crack in the sublayer, under the surface-hardened layer. This statement is not vindicated by direct results. Moreover, it contradicts the experimental data.

A queue of dislocations, which is forced against some obstacle by the applied stress, is called a dislocation pile-up. Pile-ups are important in the initiation and propagation of deformation in polycrystalline metals. The leading dislocation is blocked by a grain boundary, as shown in Figure 9.33 for screw dislocations. The leading dislocation is affected not only by the applied stress, but also by an interaction force with other dislocations in pile-up. Pile-ups form until the force on the leading dislocation is sufficient to make them break through the barrier or, for screw-dislocation pile-ups, until the force suffices to cause cross-slip of the leading dislocations [79].

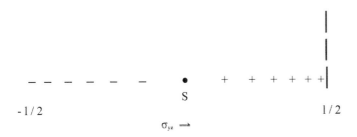

**Fig. 9.33** The double pile-up of screw dislocations is blocked by an obstacle: $l$ is the length of the pile-up, S is a source of dislocation loops. The pattern shows a section of the loop sequence in the plane of drawing; the edge components of the loops are parallel to the plane of the drawing.

In [80] the fatigue limit of aluminum alloys is modeled as the dislocation pile-ups against the grain boundaries. Dislocations move across the whole grain without being hindered until reaching a grain boundary. The grain boundary blocks the motion of a leading dislocation. As the authors believe, the fatigue limit is fundamentally simulated by adopting an analogue of fictitious fatigue microscopic crack nucleating in slip bands within a grain because the cyclic slip is essential for fatigue crack nucleation. The fatigue limit is associated with the delayed dislocation pile-ups against grain boundaries during fatigue cycling. In other words, the continuum approximation for the dislocation pile-up is assumed by the authors to be equal to a shear crack. The grain boundary is taken to be a perfect obstacle and the pile-up is modeled in terms of a continuous distribution of dislocations. Authors consider the pile-up which is subjected to a uniform stress as a fictitious crack of length $l$ in the same grain. As a result of lattice displacements the fictitious crack starts to become a real crack. The concept of the fictitious fatigue crack is not clearly defined. An analogous model has been investigated earlier quantitatively in [81].

The Paris relationship (9.3) provides an estimation of the fatigued component life [71]. (9.3) may be rewritten as

$$\frac{da}{dN} = C[Y\Delta\sigma(\pi a)^{\frac{1}{2}}]^m \qquad (9.13)$$

where $\Delta\sigma = \sigma_{max} - \sigma_{min}$. Integrating (9.13) we have

$$\int_0^{N_f} C(Y\Delta\sigma)^m \pi^{\frac{m}{2}} dN = \int_{a_n}^{a_f} \frac{da}{a^{\frac{m}{2}}} \tag{9.14}$$

where $N_f$ is the number of cycles to fracture, $a_n$ is the length of a nuclei crack, $a_f$ is the length of a crack that results in fracture.

For $m = 2$ the estimation of the fatigue life is given by

$$N_f = \frac{1}{CY^2(\Delta\sigma)^2\pi} \ln \frac{a_f}{a_n} \tag{9.15}$$

The number cycles until the fracture, in the general case ($m \neq 2$), can be expressed as

$$N_f = \frac{2}{(m-2)CY^m\Delta\sigma^m\pi^{\frac{m}{2}}} \left[ a_n^{\frac{2-m}{2}} - a_f^{\frac{2-m}{2}} \right] \tag{9.16}$$

The problem of fatigue-life prediction is vitally important in applications of contemporary superalloys. The advanced gas-turbine components are subjected to a joint action of mechanical forces and extreme heat loads. These essential factors promote thermomechanical fatigue cracking. Superalloys are used in hot sections gas turbines.

The author of [82] presented a model that was developed to predict thermomechanical crack initiation. The model estimates the rate of crack growth in superalloys. The author chooses a so-called $J$-integral as a suitable parameter. The $J$-integral is an empirical parameter that depends on the density of the strain energy during the process of the crack growth. In order for the model to fit the experimental data one cannot avoid introducing many material constants. The developed model is formulated by the author in the form of seven equations. The equations contain eight material constants to be further determined.

The crack growth rate is given by

$$\frac{da}{dN} = AJ_{eff}v^c \tag{9.17}$$

where the last factor takes into account the influence of temperature on the crack growth.

$$v = 1/\int^{cycle} \exp\left[Q_0\left(\frac{1}{T_0} - \frac{1}{T}\right)\right] dt \tag{9.18}$$

A good empirical formula for the threshold $J$-integral $J_{th}$ was found to be

$$J_{th} = J_0 \exp\left[Q_0\left(\frac{1}{T_{max}} - \frac{1}{T_J}\right)\right] \tag{9.19}$$

where $T_J$ is the temperature below which $J_{th} = J_0$. Equality (9.19) implies that $T_{max} \geq T_J$.

$$J_{eff} = \pi(a + a_0)(\beta^2 f W_{eff}^{tot})^2 - J_{th} \qquad (9.20)$$

$$W_{eff}^{tot} = \frac{(\sigma_{max} - m\sigma_{min})^2}{E_{max}} + \alpha W_p \qquad (9.21)$$

In (9.17)–(9.21) $a$ is the crack length, $N$ is the number of cycles, $J_{eff}$ is an empirical parameter that is related to the strain energy density, $J_{th} = J_0$ is the threshold integral, $\sigma_{max}$ and $\sigma_{min}$ are the maximum and minimum stresses of cycle, respectively, $W_{eff}^{tot}$ is the effective strain energy density, $E_{max}$ is the elastic modulus in the $\sigma_{max}$ direction, $W_p$ is the plastic energy density, $T_{max}$ is the maximum absolute temperature of the cycle, and $A, b, c, J_{th}, a_0, Q_0, T_0, \alpha$ are material constants.

The author estimated some material constants from tests by applying least-squares regression analysis. Specimens of a special tube shape were used for thermomechanical fatigue tests. Values of other constants were taken from more or less plausible suppositions.

The model prediction capability was tested for the uncoated cobalt-based MAR-M509 superalloy and the nickel-based coated single-crystal PWA 1480 superalloy. Figure 9.34 illustrates predictions of the model. This empirical model was able to predict crack growth up to half of the crack, that is, about 1.27 mm. As one can see from Figure 9.34 the crack grows more rapidly than the model predicts.

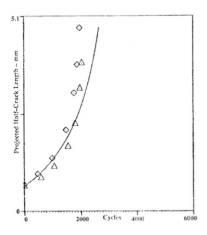 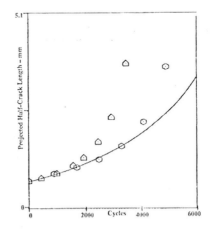

**Fig. 9.34** Model predictions of the thermomechanical crack growth for the uncoated MAR-M509 superalloy. Points are fatigue test data, curves present the model prediction (after [82]).

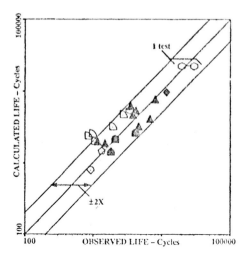

**Fig. 9.35** Correlation between predicted and test data for the coated PWA 1480 superalloy (after [82]).

The next graph in Figure 9.35 presents a comparison of prediction with test data for the overlay coated PWA 1480 superalloy. The calculated lives of the crack initiation averaged as slightly higher than the observed crack initiation lives.

As a whole, the model can capture many cracking effects such as thermal expose, single-crystal anisotropy and the coating thickness. Its application requires a large number of preliminary tests. One has to determine many constants of the material.

The described prediction technique seems to be suitable for superalloys on condition that the material, with the same homogeneous properties, is supplied evenly.

## 9.7
## Summary

Among the favorable structural factors that have a significant influence on the fatigue lifetime are grains on the nanometric scale, compressing residual macroscopic stresses, a high near-surface dislocation density, an intentional change in the chemical composition in a thin surface layer, phase transformation, and the preferred orientation of grains. These factors or a part of them are achieved specifically by severe plastic deformation in near-surface regions.

The surface strengthening processes are shot peening, treatment by ball-bearings in an ultrasonic field, deep rolling, mechanical attrition, wire-brushing, hammering and laser shock peening.

Profiles of the compressor blades, the surface of which was treated by vibratory polishing, plasma nitriding, shot peening and strengthening by ball-bearings in the ultrasonic field, were all studied. Shot peening ensures minimal values of $Ra$ and $Rz$. The average deviation in the profile does not exceed 0.27 μm and values of $R_z$ are 1.73–1.75 μm. The operation of blades in a gas turbine leads to an appreciable deterioration of the surface. The deviation in the profile is double that before operation.

Treatment technology affects the residual stresses on the surface of specimens and components. Plastic strain induces at the surface compressive residual stresses of typical value from $-450$ MPa to $-600$ MPa and up to 100–300 μm depth. Some authors report considerably larger values, to $-1200$ MPa and to 500 μm depth.

The nanostructuring treatment of the surface leads to a very marked improvement in the fatigue performance for low-alloy steels, stainless steels, titanium and aluminum-based alloys, and superalloys. For example, deep rolling of specimens of a titanium-based alloy increased the fatigue lifetime by two orders of magnitude. For some treatments the fatigue limit can be further improved by combining of the nanostructure treatment with an annealing.

The induced residual stresses are affected by the cyclic load. The stresses can redistribute and relax due to repeated cyclic loading, especially at high temperatures. However, the compressive residual stresses do not disappear completely. The considerable relaxation of induced compressive stresses during the operation of compressor blades is unlikely to occur in practice. The operation only decreases the stress depth from 80–100 to 70 μm.

Materials with high stacking fault energy have a strong propensity for cross-slip. In these materials the dislocation cell structure is formed during cycling. This process is independent of prior cold work hardening. In contrast to the microscopic stresses, residual macroscopic stresses were found to be more stable during fatigue. There is transmission electron microscope evidence that the subsurface non-nanocrystalline layer with high initial dislocation density is strongly affected by fatigue.

The approach of the elastic fracture mechanics to the problem of fatigue of materials does not take into consideration the atomic structure of the fatigued material. This approach is successful in engineering design. The main concept of elastic fracture mechanics is the stress intensity factor $K$, which is a measure of intensity of the stress field near the crack tip. The critical value of the stress intensity factor, which causes initiation of crack advance, is a function of the material microstructure, the test temperature, the loading mode, and the strain rate.

Paris assumed an empirical formula for the growth rate of a crack under the cyclic loading

$$\frac{da}{dN} = C\, \Delta K^m \qquad (9.22)$$

where $a$ is the length of the fatigue crack, $N$ is the number of cycles, $\Delta K$ is the range of the stress intensity factor, and $C$ and $m$ are fitting constants. $C$ and $m$ depend on material properties and microstructure, the mean load ratio, the loading frequency, the loading mode, the test temperature, and on the environmental influence. No physical interpretation of these constants is known. One must determine their values from fatigue tests.

According to the molecular dynamics method one has to specify the initial positions and velocities of all the atoms in a model atomic crystal. The Newton equations of motion are then integrated numerically. It was found that the vacancy lines enhanced the strain fields at the crack tip, resulting in lower rates of release of strain energy than in a defect-free crystal.

The phenomenon of fatigue is very complex. It is difficult to compare data obtained by different investigators, in a meaningful way.

From the physical point of view there are three levels of approach to the fatigue problem: macroscopic, microscopic, and atomic. In the first one the material is considered as a homogeneous continuum. The second approach operates with microscopic concepts: microstructure, microscopic crack, fracture. The atomic consideration of the problem deals with dislocation density evolution, with part of vacancies, the generation of surface steps, the distribution of macroscopic stresses, and a variation in the grain and subgrain structures. For the third approach, on the atomic scale, there is not enough experimental information. These approaches do not exclude each other. Some models have the characteristic features of several approaches.

According to a typical semi-empirical fatigue model, in order to predict the fatigue life, it is necessary to determine fatigue limits of the base material in bending as well as in torsion tests, and also to know the residual stress distribution and width of X-ray reflections.

The problem of the fatigue-life prediction is vitally important in applications for contemporary superalloys. The advanced gas-turbine components are subjected to the joint action of mechanical forces and extreme heat loads. These essential factors promote thermomechanical fatigue cracking. Superalloys are used in hot sections of gas turbines.

A model of thermomechanical crack propagation can capture many cracking effects such as thermal exposure, single-crystal anisotropy and coating thickness. In order for the model to fit experimental data one cannot avoid introducing many material constants. Some constants of the model must be determined from preliminary fatigue tests. One should take values of other constants from more or less plausible assumptions.

# 10
# The Physical Mechanism of Fatigue

It goes without saying that physicists cannot even dream of deriving laws of fatigue, either from classic physics laws, or from first principles. The phenomenon of fatigue is too complex for our today's knowledge level; "And the number of variables which can affect the fatigue behavior of a structure is large ... The physical understanding of the fatigue phenomena is essential for the evaluation of fatigue predictions" [78]. Therefore, it is appropriate to indicate possible directions for further investigation of physical mechanisms that influence fatigue phenomena.

We would like to describe the development of fatigue damage in terms of physical processes in the crystal lattice of metals and alloys. We shall try to estimate the value of specific physical parameters which determine fatigue failure. The major topics in this chapter are associated with the dislocation interaction, mechanisms of crack initiation and growth, periods of fatigue crack propagation, the role of the crystal lattice vacancies and the dependence of the crack growth rate on the stress gradient near the crack tip.

## 10.1
## Crack Initiation

Experiments on fully reversed fatigue under fixed amplitudes indicate the existence of saturation after an initial cyclic work hardening. The plastic shear strain $\gamma_{pl}$ is a function of the shear stress $\tau_s$. After a number of cycles $N_s$ the hysteresis loops saturate with a peak tensile stress of $\tau_s$ and a strain of $\gamma_s$.

Figure 10.1 illustrates the variations in the resolved shear stress as a function of the resolved shear strain. The saturation of the strain is evidence of the dislocation movement in the intersecting crystalline planes. It also indicates interaction between mobile dislocations. The density of dislocations within the persistent slip bands is maintained at an approximately constant level. Dynamic equilibrium between dislocation multiplication and annihilation is assumed to occur.

The deformation of polycrystalline alloys subjected to a cyclic load is usually characterized by cyclic $\sigma - \gamma$ curves similar to those for single crystals.

*Strained Metallic Surfaces.* Valim Levitin and Stephan Loskutov
Copyright © 2009 WILEY-VCH Verlag GmbH & Co. KGaA, Weinheim
ISBN: 978-3-527-32344-9

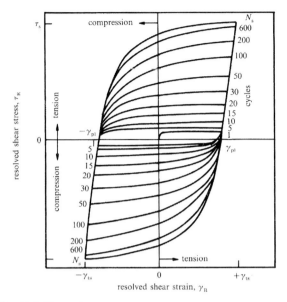

**Fig. 10.1** The typical hysteresis loops during cycling of a face-centered cubic single crystal. Reprinted from [71] with permission from Cambridge University Press.

Our investigations (Chapter 5) have determined that the movement of dislocations plays a significant role in the fatigue-failure preparation. A decrease in the electronic work function during fatigue tests provides evidence that the mobile dislocations emerge from the surface.

A force is known to attract dislocations to the surface of a specimen (Figure 10.2). A screw dislocation is drawn to the surface by the force, $F_x$, which

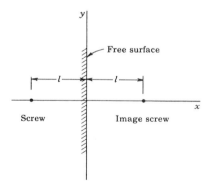

**Fig. 10.2** A screw dislocation parallel to the free surface, and an image dislocation that has the sign opposite to that of the real dislocation.

can be expressed as

$$\frac{F_x}{L} = \frac{\mu b^2}{4\pi l} \tag{10.1}$$

where $L$ is the length of the dislocation, $\mu$ is the shear modulus, $b$ is the Burgers vector and $l$ is the distance from surface.

The yield strength of a material implies a mass motion of mobile dislocations under the influence of an applied external stress. However, motion of the dislocations caused by cycling loading occurs long before the yield strength is reached.

It was shown as early as 1941 [63] that the energy required to produce a certain amount of slip inside a solid is about twice that required to produce the same amount of slip at the surface.

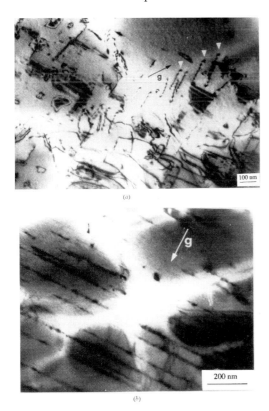

**Fig. 10.3** Transmission electron micrographs showing the dislocation structure of a superalloy after six cycles with $\gamma_{pl} = 1.1 \times 10^{-3}$: a, stacking faults and loops in the $\gamma'$-phase, and dislocation pairs in the matrix; b, the same area looking edge on. $\vec{g} = \{111\}$. Reprinted from [71] with permission from Cambridge University Press.

We have already illustrated the active development of a substructure in fatigued EP479 superalloy (Figure 9.21). The structure of the nickel-based PWA 1480 superalloy[1] is presented in Figure 10.3. Dislocations slip along planes {111}. Dislocations with the oscillating contrast that are marked by arrows were found to have the $a/2 <110>$ Burgers vector. Micrographs imply that the deformation mechanism during cycling involves the shearing of the $\gamma'$-precipitates. One can see the interaction of mobile dislocations with the $\gamma'$-precipitates in Figure 10.4. Dislocations are pressed to the edges of $\gamma'$-particles. They continue to move after entering some particles.

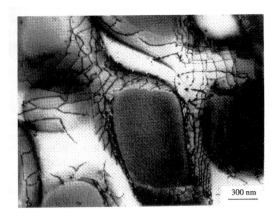

**Fig. 10.4** Dislocations by-passing particles of the $\gamma'$-phase of a superalloy after 785 cycles at temperature 1366 K (1093°C). Data of W.W.Milligan. Reprinted from [71] with permission from Cambridge University Press.

Thus, a considerable body of evidence allows us to conclude that deforming dislocations move during the cycling of alloys. The slip of the majority of dislocations is concentrated in strained metals in slip bands.

Many authors observed that fatigued metal extruded from the slip bands. Typical extrusions and intrusions appear on the polished surface of copper specimens fatigued for 1% [85] (Figure 10.5). They occur in relatively large number and in similar dimensions, along slip bands formed during cyclic deformation. A purely mechanical process of cyclic slip appears to involve the cyclic slip along crystal planes.

A scheme of a surface defect formation is shown in Figure 10.6. Here two intersecting slip bands operate sequentially during both positive and negative phases of the stress cycle. Where the sequence is maintained, an intrusion is formed on the band that first becomes active during each half-cycle, and an extrusion is formed on the other one. One band becomes active before the

---

**1)** The nominal composition of the PWA 1480 superalloy is (wt%) 10.4 Cr, 4.8 Al, 1.3 Ti, 11.9 Ta, 5.3 Co, 4.1 W, with the balance being Ni.

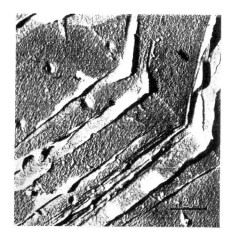

**Fig. 10.5** The electron pattern of fatigued copper showing intrusions along slip bands. Before the fatigue tests, flat specimens were annealed and then electrolytically polished.

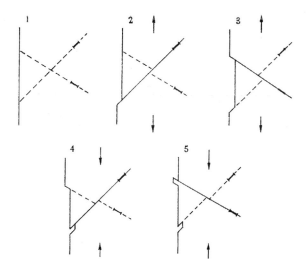

**Fig. 10.6** Scheme of the sequence of slip movements producing an extrusion and an intrusion.

other if it is more favorably oriented to the applied stress. As this stress rises to its peak during a half-cycle, this favorably oriented band becomes more completely work-hardened than the other, but nevertheless is expected to operate first on the return half-cycle because of the Bauschinger effect. The second cycle, identical with the first one, would double the length and depth of the extrusion and intrusion, respectively, but would not alter their thickness. An imperfectly repeating cycle slip, generally diminishing in amplitude produces short, thin, irregular extrusions and intrusions. A dislocation model illustrating the mechanism leading to the formation of the surface roughness [70] is shown in Figure 10.7.

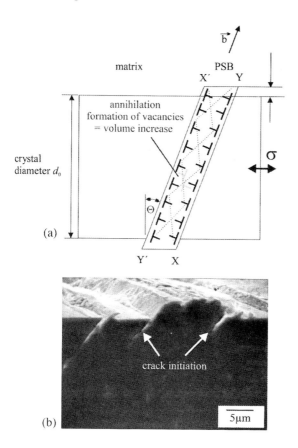

**Fig. 10.7** a, scheme of the mechanism of extrusions and intrusions at a persistent slip band (PSB) according to a model of Essmann, Göselle and Mughrabi; b, crack initiation at the persistent slip band in cyclically deformed copper, data of Ma (after [70]).

Within slip bands the annihilation of edge dislocations of opposite sign leads to the formation of vacancies or interstitials. This mechanism becomes possible because of the generation and movement of dislocations within the slip bands during alternating tension and compression periods. Annihilation of vacancy-type dislocation dipoles results in a change in the operating slip plane. The effective slip planes become slightly inclined with respect to the direction of the persistent slip band. This is represented by the line $XX'$ and $YY'$ for the load reversal in Figure 10.7. The volume, newly created by the vacancies, manifests itself in the formation of extrusions, or intrusions in the case of predominant interstitial-type annihilation. A portion of dislocations moves along dashed lines shown schematically in Figure 10.7a. Rows of interface dislocations directed in the inner slip band are formed. Assuming all these dislocations emanate at the free surface, steps and extrusions would be the consequence. Thus, surface microscopic cracks start easily in slip bands. An eventual mechanism of hollow initiation is also shown in Figure 10.8.

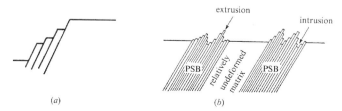

**Fig. 10.8** Slip bands at the metal surface: a, a series of steps resembling a staircase pattern produced by monotonic plastic strain; b, the rough surface consisting of hills and valleys produced by cyclic plastic strain. PSB are the persistent slip bands at the cycling surface of a specimen.

Schijve [78] notes that it is to be questioned why cyclic slip is not reversible. He believes that there are two reasons for non-reversibility. One argument is that cyclic loading implies that not all dislocations can return to their original positions. This reasonably implies that the dislocation slip is not an thermodynamically reversible process. Another important aspect is the interaction of metals with the environment. The free step surface is rapidly covered with a thin oxide layer and adsorption of foreign atoms also occurs.

We would like to emphasize that a process of accumulation of fatigue defects is in fact reversible up to some point in time. The non-destructive work function technique enables one to discover a competition of processes of generation and annihilation of defects in the crystal lattice.

Figure 10.9 illustrates this phenomenon. We have measured the electron work function during a fatigue test along 11 longitudinal lines of the specimen gauge. The lifetime of the specimen is $20 \times 10^6$ cycles. Let us note the initial stage of the cycling. The work function decreases after $2 \times 10^6$ cycles from 4.640 eV to 4.590 eV,[2] but its value reaches an initial one after $3 \times 10^6$ cycles. Consequently, surface steps, which are responsible for a decrease in the work function, can still appear and disappear. Further, one can see the next minimum of 4.600 eV and the next maximum of 4.620 eV. The height of the second maximum is less than the height of the first one. The generation of lattice defects, the emergence of dislocations, and the formation of surface

---

2) The measurement error is $\pm 0.001$ eV.

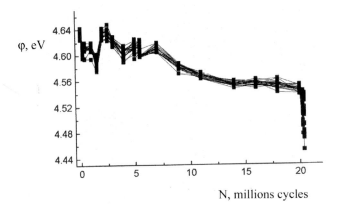

**Fig. 10.9** The work function versus the number of cycles for the EP866 superalloy. The work function values are reversible up to $2.5 \times 10^6$ cycles.

steps continues. The work function decreases gradually up to $20 \times 10^6$ cycles and then drops sharply, when fracture of the specimen occurs.

We have already seen this phenomenon in Figures 6.6 and 6.7. In the first stages of material cycling the peaks and valleys at the curves $\varphi - N$ alternate with one another.

Thus, we can observe the oscillations of the work function at the beginning of the fatigue process. Consequently, the accumulation of defects in the crystal lattice and the annihilation of defects at the beginning of cycling give way to each other. The work function measurement proved to be a unique method enabling one to observe structural processes directly during fatigue tests. This technique allows us to discover a reversibility in the beginning of a fatigue operation. The reversibility of fatigue-crack initiation during the first stage of fatigue is determined by thermodynamic factors.

The process of fatigue damage is still entirely reversible up to approximately 20–25 % of the lifetime. With this in mind, we believe that the slip bands can result in an initiation of cracks as well as in a temporary disappearance of these embryo cracks. Below we discuss the size of the fatigue crack embryo $a_c \equiv a_{incub}$, see 10.4.1.

"A valid and important conclusion is that fatigue crack initiation is a surface phenomenon" [78]. This is correct. However, the crack initiation at the surface is related to processes in the sub-surface layers as well as in the bulk.

Mura [81] worked out a mathematical model simulating the development of discrete structures in fatigued metals. Fatigue crack initiation is interpreted in the model as a state of dislocation structures, where the Gibbs free energy is the sum of the elastic strain energy, the potential energy of the applied load and the surface energy of the crack. The essence of the Mura model is the

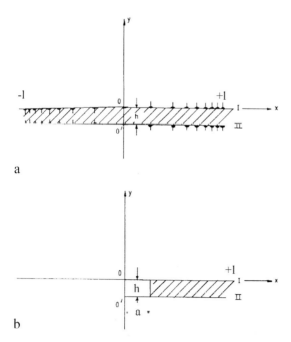

**Fig. 10.10** Two parallel slip planes I and II that are located inside a plastic strain band. $h$ is the distance between planes: a, dislocation pile-ups that are situated along slip planes store up the elastic strain energy; b, the crack initiation of length $a$ leads to a relaxation of the internal stresses and to a decrease in the energy.

conversion of dislocation pile-ups in parallel planes into the crack. The model allows for some quantitative conclusions, and this is one of its advantages.

The density of dislocation dipoles increases in a metal as a result of cyclic loading. The elastic strain energy grows with the number of loading cycles. Accumulation of the elastic strain energy can only be relieved by crack initiation along the persistent slip bands. Two parallel slip planes with dislocation pile-ups are shown in Figure 10.10.

Figure 10.11 illustrates the model of crack initiation in the slip band. The dislocation slip takes place along a favorable slip plane denoted by layer I. On this slip plane the sum of the applied stress $\tau_1$ and the dislocation stress must be equal to the threshold stress:

$$\tau_1 + \tau_1^D = \tau_{th} \qquad (10.2)$$

where $\tau_1$ is the applied stress, $\tau_1^D$ is the stress caused by the dislocation pile-up on layer I, $\tau_{th}$ is the minimum threshold stress, under which dislocations begin to slip[3]. For metals it is equal to $10^{-3}$–$10^{-2}$ of the shear modulus.

---

3) The author of [81] uses the term frictional stress instead of threshold stress.

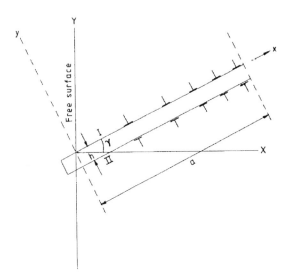

**Fig. 10.11** A slip band modeled as two parallel layers of dislocation dipole accumulation, I and II, adjoining a free surface. $a$ is the length of pile-ups, the slip plane makes an angle $\alpha$ with the normal to surface. Reprinted from [81] with permission from Elsevier.

At the first unloading a negative slip takes place on a neighboring parallel slip plane referred to the layer II in Figure 10.11. The yield condition on layer II is

$$\tau_1^D + \tau_2 + \tau_2^D = -\tau_{th} \tag{10.3}$$

where $\tau_2^D$ is the stress owing to the dislocation distribution on layer II and $\tau_2$ is the stress on the plane II. This layer is assumed to be very close to layer I and experiences the full back stress from it.

When $(\tau_1 - \tau_2) > 2\tau_{th}$ dislocations on layer II can be generated. The pile-up of negative dislocations on layer II causes a positive back-stress in layer I which generates additional dislocations on layer I in the second loading. After the second unloading, a negative slip takes place on layer II. Additional dislocations are created in layer II.

Thus, a dislocation accumulation, which alternates between layers I and II, corresponds to loading and unloading portions of the applied stress cycle. It operates as the ratcheting mechanism of the crack nucleation.

After transformations, for the total stress owing to all dislocations on layer I one can write

$$\tau_1^D + N(\tau_1 - \tau_2 - 2\tau_{th}) = 0 \tag{10.4}$$

where $N$ is the number of cycles.

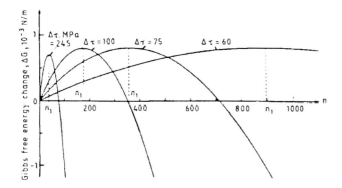

**Fig. 10.12** Variation in the Gibbs free energy with the number of loading cycles $n$. Reprinted from [81] with permission from Elsevier.

And on layer II

$$\tau_{II}^D + N(\tau_1 \quad \tau_2 \quad 2\tau_{th}) = 0 \tag{10.5}$$

where $\tau_I^D$ and $\tau_{II}^D$ are the total stresses owing to all dislocations on layer I and layer II, respectively.

Equation (10.4) then becomes an integral equation for the continuous dislocation density. Finally, the author obtains an expression for the crack length $a$ in the form

$$a = 2.3 \frac{l(1-\nu)N(\tau_1 - \tau_2 - 2\tau_{th})}{\mu} \tag{10.6}$$

where $l$ is the length of the dislocation pile-up, $\nu$ is the Poisson coefficient and $\mu$ is the shear module.

The total strain energy of dislocation pile-ups $F_1$ accumulates during alternating loading. The energy $F_2$ is released as a result of the relaxation and opening of a crack. The free energy changes in going from the state of dislocation dipole accumulation to that of a crack of size $a$. This change is given by

$$\Delta F = F_1 + F_2 - 2\gamma S \tag{10.7}$$

where $2S$ is the crack embryo area and $\gamma$ is the surface energy. The plus sign in $F_1 + F_2$ means that the system gains the elastic energy as a consequence of transformation. The minus sign in $-2\gamma S$ means that the system expends the energy to create two new faces of the crack. As a result, the increment of the free energy $\Delta F$ first increases with the number of loading cycles, then it achieves a maximum and begin to decrease.

The surface area of the critical crack embryo is determined from the equality:

$$2\gamma S = F_1 + F_2 \tag{10.8}$$

A crack embryo grows if

$$2\gamma S < F_1 + F_2 \tag{10.9}$$

The closure of the crack occurs if

$$2\gamma S \geq F_1 + F_2 \tag{10.10}$$

Inequalities (10.9) and (10.10) define the transition from the crack initiation period to the crack growth period.

The variation in the free energy with the number of loading cycles is presented in Figure 10.12. The graph shows that the more the applied load the smaller the size of the crack embryo. The critical crack length is achieved in 170 cycles under a load amplitude of 100 MPa, but under a load of 75 MPa, 350 cycles are necessary.

There is a critical number of cycles at which the free energy has a maximum and where the system becomes unstable, that is, when

$$\frac{\partial(\Delta F)}{\partial N} = 0 \tag{10.11}$$

Solving the (10.11) Mura arrives at

$$(\sigma_a \sin 2\alpha - 4\sigma_{th})N_c \approx \frac{4.6\pi\gamma}{l\zeta_1}\left(\frac{2-f}{f}\right) \tag{10.12}$$

where $\sigma_a$ is the stress amplitude, $\alpha$ is the angle of inclination of the slip plane to the surface (Figure 10.11), $\sigma_{th}$ is the threshold stress, $N_c$ is the critical number of cycles, $\gamma$ is the surface energy, $l$ is the length of the dislocation pile-up, $\zeta_1 = (h/l)^{\frac{4}{3}}$, $h$ is the distance between planes I and II (Figure 10.11) and $f$ is a factor of irreversibility of slip.

## 10.2
### Periods of Fatigue-Crack Propagation

It is important to stress that there is an initial time of fatigue when only a surface crack is initiated and the crack embryo cannot yet grow. It is preferable to term this time as an incubation period of fatigue[4].

---

[4] Preparatory processes of fatigue can be discovered by the work function technique because of surface step formations, see Chapter 5

We introduce the concept of the critical crack length, $a_c \equiv a_{incub}$ that corresponds to the critical value of cycles, $N_c \equiv N_{incub}$. If the number of cycles $N < N_{incub}$ the crack does not grow. If the number of the loading cycles $N \geq N_{incub}$ the crack length begins to increase.

It is appropriate to divide the total number of cycles to the fracture $N_f$ into three terms:

$$N_f = N_{incub} + N_{gr} + N_{inst} \tag{10.13}$$

where $N_{incub}$ is the number of cycles when the crack length is less than the critical size, the crack does not grow, and the process is still reversible (an incubation period); $N_{gr}$ is the number of cycles when the fatigue crack grows in length (a period of growth); $N_{inst}$ corresponds to the instant fracture, when the cross-section of a specimen or a component has already decreased due to the crack (a period of fracture). Generally, $N_{inst} \ll (N_{incub} + N_{gr})$.

The higher the applied stress amplitude the smaller the critical crack size. At large $\sigma_a$ $N_{incub} \ll N_{gr}$. With this condition the initiation of a crack is a very fast process, and the crack begin to grow immediately.

On the contrary, under a relatively small applied stress amplitude $\sigma_a$ the incubation period takes a considerable part of the specimen life. The number of cycles until specimen fracture becomes very large and sometimes does not occur.

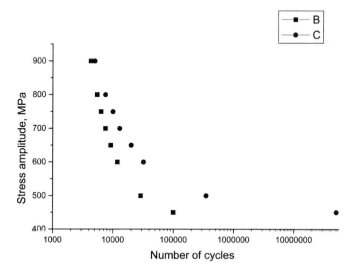

**Fig. 10.13** Effect of the stress amplitude on the number of cycles to beginning of crack growth and to the fracture of the specimen. C, number of cycles to fracture $N_f$ for the titanium-based alloy, test data of [61]; B, the calculated number of cycles during the "incubation" period $N_{incub}$ for the same alloy.

In Figure 10.13 two curves are presented: the experimental dependence of $N_f$ on $\sigma_a$, and the stress dependence of $N_{incub}$, which was calculated by us according to (10.12). Reasonable values for the constants were assumed as follows: $\mu - 3.58 \times 10^4$ MPa for a titanium-based alloy, $\sigma_{lh} = 3 \times 10^{-3} \mu$, $\alpha = \pi/4$, $\gamma = 1.885$ J m$^{-2}$ [76], $f = 0.5$, $l = 100$ μm, $h = 1.6$ nm [81].

It follows from Figure 10.13 that the incubation period lasts for $10^5$ cycles under a relatively low stress amplitude of 450 MPa, whereas the period of the crack growth continues up to $5 \times 10^7$ cycles. Consequently, under this condition $N_{incub}$ is two orders greater than $N_{gr}$. Under a stress amplitude of 900 MPa the short incubation period $N_{incub} = 4300$ cycles gives place at once to the crack growth period and the fracture of the specimen occurs after a further 5000 cycles.

Figure 10.14 illustrates the relation between periods of fatigue-crack propagation for the titanium-based alloy and for stainless steel. One can again see that the smaller the stress amplitude, the longer the incubation period of fatigue. The crack embryo can exist as long as $2 \times 10^5$ cycles, curve B Figure 10.14.

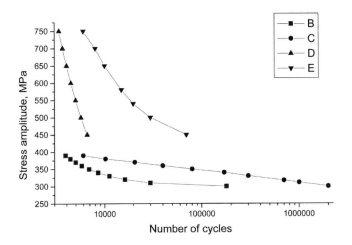

**Fig. 10.14** Dependence of the incubation period, and of the crack-growth period on the stress amplitude for two materials: C, the curve of the fatigue fracture (S – N curve) for 316L stainless steel, test data [62]; B, calculated values of $N_{incub}$ for the same steel; E, the curve of fracture for the Ti+6Al+4V alloy, test data [63]; D, calculated values of $N_{incub}$ for the same alloy.

We shall discuss the dependence of the crack embryo size on the stress amplitude in the next section.

## 10.3
## Crack Growth

The authors of [86] reported an investigation in fatigue-crack behavior at fastener holes in a low carbon steel before and after cold-expansion and cold-stretching processes. In Figure 10.15 the dependence of the crack length on the number of cycles is presented. Figure 10.15 shows that crack initiation and propagation in cold-worked specimens are significantly retarded and the crack growth rates are smaller compared to those in non-cold-worked specimens.

The fatigue crack originates at the surface, then the crack embryo begins to grow. In our opinion, a movement of vacancies to the tip of the crack plays a dominant role in crack growth.

Two questions arise. Why does an increased concentration of vacancies exist in fatigued metals in bulk and near to the surface? How can vacancies diffuse to the crack, especially at room temperature?

Dislocations move under the influence of loading in intersecting slip planes. Jogs are generated at the intersection of screw components of mobile dislocations (Figure 10.16). A jog is a segment of dislocation, which does not lie in the slip plane. The jog cannot move without the generation of vacancies. The interstitial-producing jogs are practically immobile because the energy of formation of the interstitial atoms is several orders greater than that of vacancies. Thus, the slip of the jogged dislocation components plays an important role in the increase of vacancy concentration.

There are other sources of vacancies. They are also generated because of the annihilation of edge dislocation segments of opposite sign. The concentration of vacancies produced within slip bands is larger than that in the vicinity of the free surface. The gradient of the vacancy concentration creates a driving force for the diffusion of vacancies within the slip bands to the free surface. The activation energy of diffusion along dislocation pipes is far less than that in the bulk.

One would expect a sufficient time to be available for vacancy diffusion even at room temperature since fatigue involves a sufficiently long period of time for crack initiation and propagation.

An energy $E_v$ (J at$^{-1}$) is necessary in order to create a vacancy in the bulk. There is an equilibrium concentration of vacancies in the crystal lattice of the metal. This concentration depends essentially upon the temperature $T$. One can estimate the fraction of lattice nodes in the crystal lattice $c_0$ that are occupied by vacancies as

$$c_0 = \exp\left(-\frac{U_v}{kT}\right) \tag{10.14}$$

where $k$ is the Boltzmann constant.

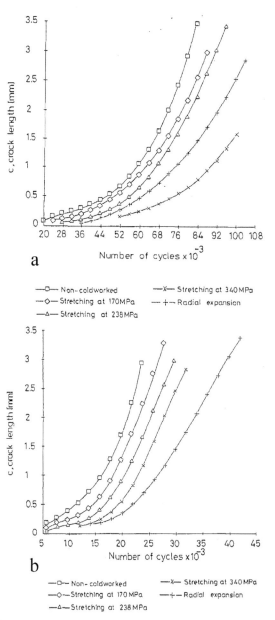

**Fig. 10.15** Experimental crack length versus number of test cycles. Amplitudes of loading are: a, 170 MPa; b, 255 MPa. Frequency of loading is equal to 23 Hz. Specimens were subjected first to various cold-working processes. Reprinted from [86] with permission from Elsevier.

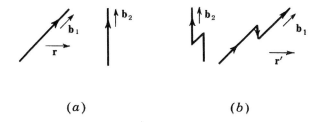

**Fig. 10.16** The intersection of two screw dislocations with Burgers vectors $\vec{b_1}$, $\vec{b_2}$, respectively. $\vec{r}$ and $\vec{r'}$ are vectors showing the direction of the movement: a, before intersections; b, after intersections. Jogs are formed at both dislocations.

The relative number of vacancies which are formed at the intersection of dislocations $c_{in}$ for face-centered crystals is given by [21]

$$c_{in} \approx 1 \times 10^{-4} \varepsilon^2 \tag{10.15}$$

where $\varepsilon$ is the strain.

By estimation, $c_{in} \gg c_0$. The number of vacancies, which are generated by the mobile intersecting screw components of dislocations, is much greater than the equilibrium number of vacancies that are formed spontaneously.

Figure 10.17 illustrates fluxes of the vacancies in a specimen subjected to the alternating tensile – compressive stresses. Vacancies near the crack are attracted by the crack tip during the tensile phase of cycling.

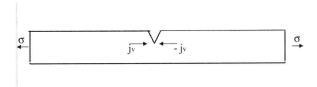

**Fig. 10.17** A specimen in the field of the cycling stresses $\sigma$ during the fatigue test. $\vec{j_v}$ and $-\vec{j_v}$ show fluxes of vacancies in the line of the crack tip.

The stress distribution near a crack tip is presented schematically in Figure 10.18. The value of the stress gradient at the tip,

$$\text{grad } \sigma = \left(\frac{d\sigma}{dx}\right)_{x=0} \tag{10.16}$$

is of great importance.

Vacancies move in the crystal lattice with a velocity of $V_v$. There is no need for vacancies to diffuse over relatively long distances. A relay-like motion of

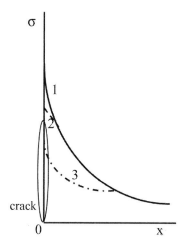

**Fig. 10.18** Stresses versus the distance from the crack tip: 1, the tensile stress near the tip of the growing crack; 2, a decrease in the maximum tensile stress in front of the tip due to a plastic strain and relaxation; 3, the drop in the tensile stress due to induced compressive residual stresses after a preliminary treatment, such as shot peening.

vacancies can occur. The growing crack absorbs the nearest vacancies as this process leads to a decrease in the free energy of the thermodynamic system crack – vacancies. The increased gradient of the vacancies leads to a displacement of the next vacancies in the direction of the crack. As a consequence, the crack that absorbs vacancies grows under the influence of alternating loading. Intersecting dislocations go on to generate vacancies.

Thus, the displacement of vacancies to the crack tip occurs due to the gradient of tensile stresses. Moreover, the falling of vacancies into the crack produces a gradient of concentration near the crack tip. For its part, the gradient of the vacancy concentration also facilitates vacancy movement in the line of the crack.

Consequently, external cycling stress creates a field of tensile stresses at the crack tip. The gradient in the tensile stress and in the vacancy concentration insure a driving force for the diffusion of vacancies in the direction of the crack tip. Vacancies are attracted to the tension area near the crack tip. The displacement of vacancies is realized in this field in the direction of the tensile stresses. If a vacancy reaches the crack, it falls down into it. The crack is a trap for the flow of vacancies. The vacancy that replaces an atom results in the breaking of interatomic bonds at the tip of the crack. This process is not reversible, of course. Figure 10.19 illustrates the process.

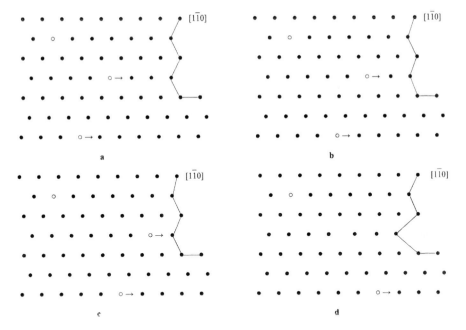

**Fig. 10.19** The scheme of vacancy diffusion to the crack tip in the field of tensile stresses: a, b, c, the sequence of vacancy displacements in the crystal lattice (shown by arrows); d, the vacancy has fallen into the crack.

The replacement of a bonding atom by one vacancy results in the increment of the crack length by one interatomic distance $b$, that is, by the module of the Burgers vector; $dn$ vacancies increase the length of the crack by $da$. Consequently,

$$da = b \; dn \tag{10.17}$$

Differentiating (10.17) we obtain the rate of crack growth

$$\frac{da}{dt} = b \frac{dn}{dt} \tag{10.18}$$

The number of vacancies through a cross-section $S$ during a time interval $dt$ is proportional to the gradient of stresses $grad\,\sigma$ and values of $S$, and $dt$:

$$dn \sim \frac{d\sigma}{dx} S \, dt \tag{10.19}$$

The number of vacancies, which approaches the crack tip, is also directly proportional to the velocity of vacancies and inversely proportional to the activation energy of their generation.

Using the dimensional method and making reasonable physical assumptions, we arrive at the formula

$$dn = \frac{V_v}{U_v} a \frac{d\sigma}{dx} S\, dt \qquad (10.20)$$

where $a$ is the length of the crack, $U_v$ is the activation energy of the vacancy generation, $V_v$ is the velocity of vacancies.

Differentiating (10.20) and combining with (10.18) we obtain an expression for the rate of fatigue-crack growth in the form

$$\frac{da}{dt} = \frac{bSV_v}{U_v} a \frac{d\sigma}{dx} \qquad (10.21)$$

We should note that (10.21) does not contain any arbitrary coefficients. Only values which have a certain physical meaning determine the rate of fatigue-crack growth.

From (10.21) we obtain

$$\frac{da}{dN} = \frac{bV_v}{\nu U_v} g S a \qquad (10.22)$$

where $g = d\sigma/dx = \mathrm{grad}\,\sigma$ and $\nu = dN/dt$ is the frequency of cycling.

Equations (10.21) and (10.22) are applicable if the crack size $a \geq a_{incub}$. In order to estimate some parameters we should turn to experimental data.

Denoting

$$G = \frac{bSV_v}{U_v} \frac{d\sigma}{dx} \qquad (10.23)$$

we obtain

$$\frac{da}{dt} = G\,a \qquad (10.24)$$

Separating the variables and integrating (10.24) we arrive at

$$\int_0^a \frac{da}{a} = \int_0^t G\,dt \qquad (10.25)$$

and

$$a = \exp(Gt) \qquad (10.26)$$

Equation (10.26) predicts the exponential dependence of the fatigue-crack length on time. On this assumption the general equation $a = f(t)$ can be expressed as

$$a = A\,\exp(Gt) \qquad (10.27)$$

where $A$ and $G$ are constants. It is appropriate to rewrite (10.27) for calculation as

$$\ln a = \ln A + Gt \tag{10.28}$$

In Figure 10.20 one can see the results of the data treatment for non-cold-worked steel and also for plastically deformed steel. The exponential dependence between the crack length and time of fatigue tests is confirmed.

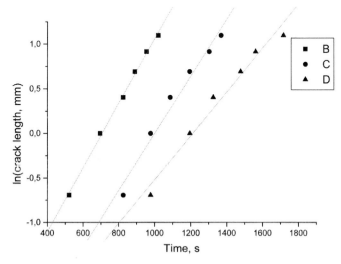

**Fig. 10.20** Dependence of the logarithm of the crack length on the fatigue test time: B, non-cold-worked steel; C, preliminary stretching at 340 MPa; D, preliminary radial expansion. The frequency of tests for low-carbon steel is 23 Hz, the amplitude is 255 MPa. Test data from [86].

Calculated values of coefficients $G$ and $A$ are presented in Table 10.1. The exponent $G$ in (10.27) that strongly affects the rate of crack growth is 1.46 times less for the cold-worked specimen than for the non-cold-worked one. There is good reason to believe that it is related to a decrease in the gradient of stresses $d\sigma/dx$ near the crack, see Figure 10.18, curve 3. The other values in (10.23) are the same for the cold strained state.

Figure 10.21 presents a comparison of the experimental dependence crack length – test time and the results of calculations on the assumption of exponential dependence between these values. The constants are presented in Table 10.1. Calculated curves fit experimental data well for the non-treated material and satisfactorily for the pre-strengthened steel.

Cold deformation of the steel results in a displacement of curves $a(t)$ to the right. The same value of the crack length is achieved at a later time. The time of crack growth to the same length increases by approximately 1.8 times.

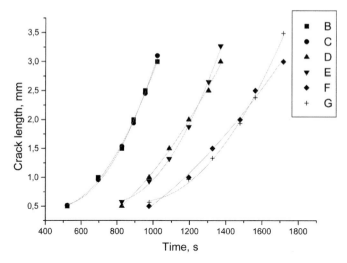

**Fig. 10.21** The length of the crack versus the time of the fatigue test: B, test data for low carbon non-cold-worked steel; C, calculated exponential dependence for the same material as B; D, data for a specimen pre-stretched at 340 MPa; E, calculated exponential dependence for the same material as D; F, data for the radially pre-stretched specimen; G, calculated exponential dependence for the same material as F. B, D, and F are the test data from [86].

**Table 10.1** The data of calculation of the coefficients in (10.27) for the fatigue-crack length $a$ for a low-carbon steel.

| Processing | A (mm) | G ($s^{-1}$) | Curves in Figure 10.21 |
|---|---|---|---|
| Non-cold-worked | 0.078 | 0.00360 | B, C |
| Stretching at 340 MPa | 0.041 | 0.00320 | D, E |
| Radial expansion | 0.051 | 0.00246 | F, G |

The residual compressive stresses cause a decrease in grad $\sigma$ near the crack tip and enhance the fatigue life of the material. The residual stress at the surface reduces the applied external stress. The strengthening surface treatment retards the growth of cracks.

In Figure 10.22 the rate of crack growth is presented as a function of the crack length. The longer the crack, the faster it grows. An acceleration in the fatigue-crack growth is known to be dangerous. The preliminary cold strain leads to a decrease in the growth rate. It is easy to see this by comparing experimental curves $B$ and $D$ in Figure 10.22.

It is important that the compressive residual stresses do not only reduce the rate of fatigue-crack growth: the compressive residual stress does more than reduce $da/dt$; it decreases the increment of the rate. Notice in Figure 10.22 that

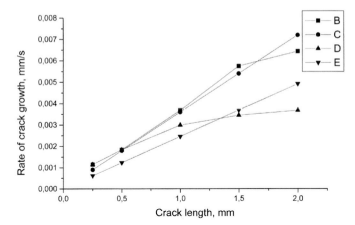

**Fig. 10.22** The rate of crack growth versus the length of the crack: B, experimental data for the non-cold-worked steel; C, calculated dependence according to (10.24) for the same material as B; D, experimental data for the radially pre-cold-stretched steel; E, calculated dependence for the same material as D.

the slope of the curves D and E to the Ox axis is less than the slope of curves B and C.

Equation (10.27) results in

$$\frac{d^2 a}{dt^2} = AG^2 \exp(Gt) \tag{10.29}$$

Thus, for a pre-cold-treated surface the increment in the crack growth rate is found to be $A_2 G_2^2 / A_1 G_1^2 = 3.28$ times lower.

Let us estimate the parameters of the crack growth rate proceeding from (10.21).

The energy of vacancy generation for $\alpha$-Fe is $U_v = 1.20$ eV at$^{-1}$ or $U_v = 1.92 \times 10^{-19}$ J at$^{-1}$ [21, 94]. The vacancy velocity is assumed to be equal to one interatomic distance per cycle, so $V_v = b\,v$. $V_v = 5.70 \times 10^{-9}$ m cycle$^{-1}$. The Burgers vector $b = 0.248$ nm;

The gradient of the stresses near the crack is assumed to be close to the gradient of the residual stresses near the surface. According to the data of the present authors and also to the results of [58, 59, 63, 65] the value of

$$\frac{|\sigma_{max}| - \sigma_{min}}{\Delta h}$$

varies from 3.1 to 6.3 MPa μm$^{-1}$. We assume $d\sigma/dx = 4.0$ MPa μm$^{-1}$.

Substituting these values into (10.21) for experimental data $da/dt = 4.60$ μm s$^{-1}$ and $a = 0.4 \times 10^{-3}$ m, we obtain $S = 3.61 \times 10^{-17}$ m$^{-2}$ and

$\sqrt{S} = 6.01 \times 10^{-9}$ m = 6.01 nm. This value appears to be of the order of the crack-tip size. It means that the needle-like tip of the crack equals 24 interatomic distances.

The authors of [67] examined the effect of residual macrostresses on fatigue-crack propagation in 1080 steel. This construction steel contains (nominal composition, mass %) 0.8C + 0.75 Mn + 0.04 P max + 0.05 S max. The residual stresses were introduced into polished double-edge notched specimens by pre-straining and press-fitting operations.

In Figure 10.23 the dependence $\ln a - t$ for the 1080 steel is shown. One can see the satisfactory fit of experimental points and the results of the calculation, especially for the non-cold-worked material.

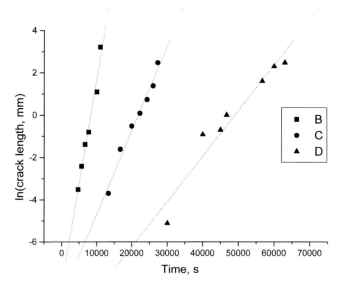

**Fig. 10.23** Dependence of the logarithm of the crack length on time: B, specimen with the tensile residual stresses of +360 MPa; C, non-cold-worked specimen; D, the specimen with compressive residual stresses after the press-fit. Results were calculated from the test data of [67].

Increasing the exponent coefficient $G$ causes a faster growth of the fatigue crack. $G$ decreases as a result of the compressive residual stresses by 1.82 times (Table 10.2, compare the second and the third lines). On the contrary, this coefficient increases as a consequence of induced tensile residual stresses by 2.45 times (Table 10.2, the first line). These changes in the $G$ value indicate that macroscopic residual stresses affect the stress gradient near the tip of the fatigue crack.

The logarithmic dependence of the rate of fatigue-crack growth on the cycle number for aluminum-based alloys[5] is illustrated in Figure 10.25.

5) The composition of alloys was not reported.

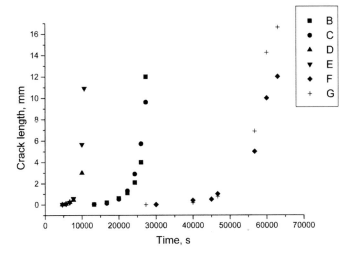

**Fig. 10.24** Crack length as a function of time during fatigue tests of 1080 steel: B and C, the non-cold-worked steel; D and E steel with tensile residual stresses of +360 MPa; F and G steel after press-fit with compressive residual stresses of −360 MPa; B, D, and F are data of measurements; C, E, and G are data of the exponential dependence (see constants in Table 10.2). Results were calculated from the test data of [67].

**Table 10.2** The data for calculation of coefficients in the equation $a = A\ \exp(Gt)$ for 1080 steel; $a$ is the fatigue crack length versus the fatigue test time.

| Processing | A (mm) | G (s$^{-1}$) | Curves in Figure 10.24 |
|---|---|---|---|
| Residual stresses +360 MPa | 0.000314 | 0.000980 | B, C |
| Non-cold-worked | 0.000174 | 0.000400 | D, E |
| Press-fit | 0.000026 | 0.000220 | F, G |

It is certain that the exponential dependence of the crack length on the cycle number holds for different materials. The rate of the crack growth is proportional to the crack length, $da/dN \sim a$. The exponential dependence between $a$ and $N$, $a$ and $t$ is related to this determined fact.

## 10.4
## Evolution of Fatigue Failure

Energy from an external source that is injected into a specimen during fatigue tests or into a component during operation is spent by several channels:
- the heat energy, that is, the energy of non-ordered vibrations of atoms;
- the elastic strain;

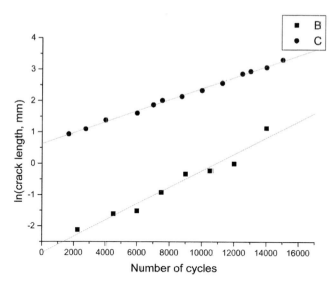

**Fig. 10.25** Logarithm of the fatigue crack length as a function of cycle number during tests of aluminum-based alloys: B, 2024 aluminum alloy, test data [61]; C, 2024-T351 aluminum alloy, test data Liu [87].

- the plastic strain, that is, the generation, motion, and interaction of dislocations;
- the initiation of a fatigue crack at the surface and its growth.

The oscillations of a specimen (Figure 10.17) can be considered as vibrations of a spring pendulum. The equation of vibrations is $x = A \sin 2\pi \nu t$, where $A$ is the amplitude, $\nu$ is the frequency, and $t$ is time. The total energy of the vibrations is known to be equal to $E = \frac{1}{2} k_e A^2$, where $k_e$ is the coefficient of elasticity of the spring. The energy that is injected into the specimen by the external source in our case is given by

$$E = \frac{1}{2} k_e A^2 \tag{10.30}$$

Note that the energy is proportional to the amplitude squared.

Let us consider, for the sake of simplicity the fatigue crack during deformation of the mode I (tensile strain), then (9.1) and (9.2) can be combined to obtain the energy release rate, $J\,m^{-2}$

$$\frac{dF}{dS} = \frac{1 - \nu^2}{E} K_I^2 \tag{10.31}$$

If we deal with the start of surface-crack growth we should consider the threshold K-value $K_{th}$. The length of the crack embryo $a_{incub}$ corresponds to $K_{th}$. This conformity seems to be a missing link between the atomic approach and elastic fracture mechanics.

If we note that $K_I = 1.12\sigma\sqrt{\pi a}$ (Figure 9.28b) we obtain

$$K_{th} = 1.12\sigma_a\sqrt{\pi a_{incub}} \qquad (10.32)$$

Substituting (10.32) into (10.31) we find

$$\frac{dF}{dS} = \frac{3.94(1-\nu^2)}{E}(\sigma_a)^2 a_{incub} \qquad (10.33)$$

Keeping in mind that $dF/dS$ is, by definition, the energy, which is spent to create two surfaces of the crack, so $dF/dS = 2\gamma$ we arrive at the equation for the length of the fatigue-crack embryo

$$a_{incub} = \frac{0.508\gamma E}{(1-\nu^2)(\sigma_a)^2} \qquad (10.34)$$

Consequently, the length of the crack embryo that can grow is inversely proportional to the stress amplitude squared.

Figure 10.26 illustrates the effect of the applied stress amplitude on the length of the crack embryo for five materials. The composition of alloys, values of Young's moduli [96] and the surface energies [97] are presented in Table 10.3.

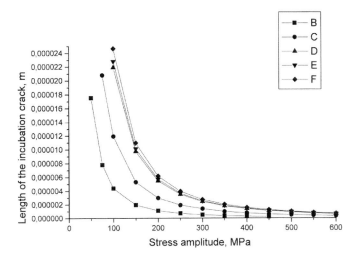

**Fig. 10.26** The length of an incubation crack versus the stress amplitude: B, an Al-based alloy; C, a Ti-based alloy; D, a low-alloy steel; E, a superalloy; F, a stainless steel. See Table 10.3 for details.

It is obvious from curves that, under stress amplitudes of more than 300 MPa, the length $a_{incub}$ is of the order of 1–2 μm for Al- and Ti-based alloys. The crack begin to grow almost at once in non-nanostructured materials. Other alloys give the $a_{incub}$ value of 4–5 μm and crack growth occurs later.

**Table 10.3** The Young modulus $E$ and the surface energy $\gamma$ for some alloys.

| Alloy | Composition (wt.%) | E (GPa) | $\gamma$ (J m$^{-2}$) |
|---|---|---|---|
| Al-based | Al+4.4Cu+1.5Mg+0.6Mn +0.5Si+0.5Fe | 73 | 1.075 |
| Ti-based Ti metal 811 | Ti+9Al+1Mo+1V | 124 | 1.690 |
| Fe-based Low alloy steel | Fe+0.16C+0.9Mn+0.55Ni + 0.50Cr+0.20Mo | 208 | 1.900 |
| Superalloy Nimonic 263 | Ni+20Cr+20Co+6Mo +2Ti+0.7Fe+0.5Al | 224 | 1.810 |
| Stainless steel AISI 17-7 PH | Fe+0.09C+17Cr+7.1Ni + 1Mn+1Al+1Si | 204 | 2.150 |

For steels and superalloys under stress a amplitude of from 150 to 300 MPa, the embryo length is found to be 4–10 μm. The values of $N_{incub}$ and $N_f$ increase correspondingly in comparison with Al- and Ti-based alloys. The stress amplitude from 100 to 150 MPa for these alloys does not lead to crack-embryo propagation. As we should expect Al-based alloys have a lower fatigue strength than steel, because under the same stress amplitude their $a_{incub}$ is several times less than that of steel.

We have discovered that the value of $grad\ \sigma$ at the tip of the growing crack decreases during fatigue tests.

From (10.22) we obtain

$$\frac{g_n}{g_1} = \frac{a_1}{a_n} \cdot \frac{da_n/dN_n}{da_1/dN_1} \qquad (10.35)$$

where subscripts 1 and $n$ denote the first and the $n$-measurements of the crack length, respectively. The length of fatigue cracks and the rate of crack growth have been measured [66] for steel. We have calculated relations $g_n/g_1$ according to these data. The results are illustrated in Figure 10.27.

A rectilinear dependence is observed for the non-treated specimen as well as for the specimen with the strengthened surface. The curve C is located above the curve B since the same crack length is achieved later.

One can see from Figure 10.28 that a decrease in the stress gradient depends mainly on the length of the crack. The drop in the gradient is explained by the plastic strain that takes place in front of the growing crack.

The equation for the curve B in Figure 10.28 is

$$\frac{g_n}{g_1} = 0.832 - 0.112a \qquad (10.36)$$

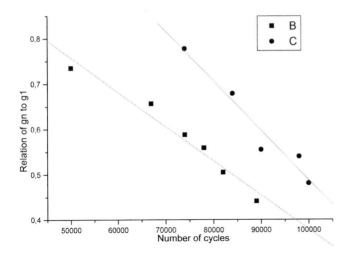

**Fig. 10.27** Effect of the number of cycles on the stress gradient at the crack tip: B, non-treated specimen; C, radially pre-stretched specimen.

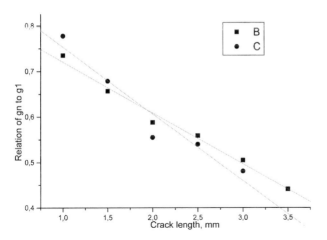

**Fig. 10.28** The stress gradient at the crack tip versus the crack length: B, non-treated specimen; C, radially pre-stretched specimen.

The equation for the curve C is[6]

$$\frac{g_n}{g_1} = 0.900 - 0.147a \tag{10.37}$$

The stress gradient drops almost twice as much due to the plastic strain near the growing fatigue crack.

6) In these empirical equations the crack length $a$ is given in mm, coefficients 0.112 and 0.147 are given in mm$^{-1}$.

We would like also to find the dependence of the number of cycles to fraction $N_f$ on the length of the fatigue crack $a_f$. Let us return to (10.22):

$$\frac{da}{dN} = P \cdot a \cdot S \cdot g \tag{10.38}$$

where

$$P = bV_v/vU_v \tag{10.39}$$

From experiment we have obtained, for non-treated steel[7)]

$$\frac{g_N}{g_0} = W - Ma \tag{10.40}$$

where the constants $M$ and $W$ are determined from dependence $g_N/g_0 - N$, see Figure 10.27.

Substituting (10.40) into (10.38) we arrive at

$$\frac{da}{dN} = P \cdot a \cdot g_0 \cdot S(W - Ma) \tag{10.41}$$

Thus we can write

$$\int_0^{N_f} dN = \int_{a_{incub}}^{a_f} \frac{1}{P \cdot g_0 \cdot S} \cdot \frac{da}{a(W - Ma)} \tag{10.42}$$

Integrating we find finally

$$N_f = \frac{1}{P \cdot W \cdot g_0 \cdot S} \ln \left| \frac{a_f(W - Ma_{incub})}{a_{incub}(W - Ma_f)} \right| \tag{10.43}$$

It is of interest to estimate values of $g_0$ and $S$. The following reasonable values may be assumed to be equal to: $b = 0.248$ nm; $V_v = 5.704 \times 10^{-9}$ m s$^{-1}$ (one vacancy jump per cycle of the stress vibration); $v = 23$ Hz; $g_0 = 3 \times 10^{12}$ or $g_0 = 4 \times 10^{13}$ Pa m$^{-1}$; $W = 1$; $M = -0.11$ mm$^{-1}$. In this case $P = bV_v/vU_v = 0.3203$ m N$^{-1}$.

We have calculated values of $\sqrt{S}$ for four ratios of $a_f/a_{incub}$, namely $10^2$, $10^3$, $10^6$ and $10^9$ and for two given values of $g_0$, namely $4 \times 10^{12}$ and $4 \times 10^{13}$ Pa m$^{-1}$. The results are presented in Table 10.4.

The tip of the fatigue crack looks like a needle point. The value of the crack tip varies from 2.4 to 40.9 nm if the ratio $a_f/a_{incub}$ is assumed to change by seven orders of magnitude. This means that the tip of the crack is estimated to be of the order of 10–170 interatomic distances. Approximately from 10 to 170 interatomic bonds are destroyed simultaneously in front of the growing crack.

---

7) Replacing subscript notations $n \to N, 1 \to 0$.

**Table 10.4** Estimation of values that determine fatigue-crack growth.

| $N_f$ | $a_f/a_{incub}$ | $g_0 \cdot S$ ($10^{-3}$ N m$^{-1}$) | $g_0$ (Pa m$^{-1}$) | $\sqrt{S}$ (nm) |
|---|---|---|---|---|
| $10^4$ | $10^2$ | 2.14 | $4 \times 10^{12}$ | 23.1 |
| | | | $4 \times 10^{13}$ | 7.3 |
| | $10^3$ | 2.86 | $4 \times 10^{12}$ | 26.7 |
| | | | $4 \times 10^{13}$ | 8.5 |
| | $10^6$ | 5.02 | $4 \times 10^{12}$ | 35.4 |
| | | | $4 \times 10^{13}$ | 11.2 |
| | $10^9$ | 7.18 | $4 \times 10^{12}$ | 42.4 |
| | | | $4 \times 10^{13}$ | 13.4 |
| $10^5$ | $10^2$ | 2.14 | $4 \times 10^{12}$ | 7.3 |
| | | | $4 \times 10^{13}$ | 2.3 |
| | $10^3$ | 2.86 | $4 \times 10^{12}$ | 8.4 |
| | | | $4 \times 10^{13}$ | 2.7 |
| | $10^6$ | 5.02 | $4 \times 10^{12}$ | 11.2 |
| | | | $4 \times 10^{13}$ | 3.5 |
| | $10^9$ | 7.18 | $4 \times 10^{12}$ | 13.4 |
| | | | $4 \times 10^{13}$ | 4.2 |

One can simplify (10.43). Since $W = 1$ (before tests $N = 0, g_N = g_0$) we obtain

$$N_f = \frac{1}{P g_0 S} \ln \left| \frac{a_f}{a_{incub}(1 - Ma)} \right| \quad (10.44)$$

where $M$ is of the order of 0.1 mm$^{-1}$.

Up to this point we have considered the velocity of vacancies as one interatomic jump during one cycle of the applied stress, that is $V_v = b\nu$ m s$^{-1}$. In general case the number of jumps during the cycle can be dependent on the stress amplitude. In this case

$$V_v = \Gamma b \nu \quad (10.45)$$

where $\Gamma$ is a mean number of jumps for the cycle. Then $\nu \Gamma$ is the number of interatomic jumps during one second.

Substituting (10.39) and (10.45) into (10.43) we find

$$\Gamma g_0 S = \frac{U_v}{b^2 N_f} \ln \left| \frac{u_f(W - Mu_{incub})}{a_{incub}(W - Ma_f)} \right| \quad (10.46)$$

Remind that the gradient $g_0$ as well as number of vacancy jumps $\Gamma$ both depend on the stress amplitude $\sigma_a$. It is unlikely that one could determine these values each separately.

## 10 The Physical Mechanism of Fatigue

The values were assumed to be equal for aluminum-based alloys to: $b = 0.286$ nm; $U_v = 1.216 \times 10^{-19}$ J at$^{-1}$ [88]; $M = 0.1$ mm$^{-1}$; $a_f = 8$ mm; $a_{incub} = 0.5 - 1.2$ µm (Figure 10.26).

Figure 10.29 illustrates the effect of the stress amplitude on the value of $\Gamma g S$. The linear dependence between $\ln(\Gamma g S)$ and $\sigma_a$ is in accordance with the function

$$\Gamma g S = A \exp\left(\frac{\sigma_a}{\sigma_{th}}\right)$$

where $A = 3.04 \times 10^{-8}$ N m$^{-1}$ and $\sigma_{th} = 31.1$ MPa. The linear dependence takes place in the interval from 237 to 300 MPa. For a stress amplitude of less than 237 MPa, points are located at the graph below the straight line.

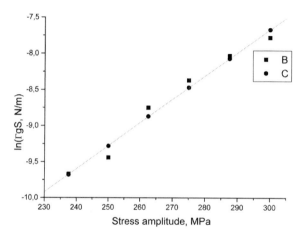

**Fig. 10.29** Effect of the number of cycles on the logarithm of $\Gamma g S$ for non-cold-worked aluminum-based alloy. B, experimental data; C, exponential dependence.

The same linear dependence holds for specimens that were subjected to short peening (Figure 10.30). Here the linearity takes place for stress amplitude from 300 to 340 MPa. The constants $A$ and $\sigma_{th}$ are equal to $7.2 \times 10^{-7}$ N m$^{-1}$ and 79.9 MPa, respectively. The value of $\sigma_{th}$ more than doubles as a result of shot peening.

The greater the stress amplitude the greater stress gradient near the crack tip and the greater number of vacancy jumps during the cycle. At relatively small applied stresses their product increases slowly. At relatively large amplitudes the value $\Gamma g S$ increases exponentially, which means an exponential rate of the crack growth.

Shot peening improves the fatigue strength. Compare, for instance, the data for 300 MPa in Figures 10.29 and 10.30. The products $\Gamma g S$ are equal to $4.53 \times$

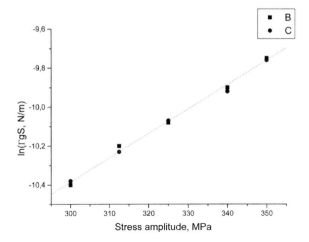

**Fig. 10.30** As in Figure 10.29 for short-peened aluminum-based alloy. B, experimental data; C, exponential dependence.

$10^{-4}$ and $3.04 \times 10^{-5}$ N m$^{-1}$, respectively. The product of the stress gradient and the number of vacancy jumps decreases by 15 times as the result of surface shot peening.

## 10.5
## S – N curves

Graphs that are plotted in the coordinates stress amplitude $\sigma_a$ – number of cycles to fracture $N_f$ are termed $S - N$ curves.

Numerous fatigue tests on specimens have been carried out for an estimation of the reliability of components. Unnotched and notched specimens are used, and the data are usually plotted on a semi-logarithmic scale. This type of dependence is presented in Figures 9.9–9.12.

One usually has available data obtained under time-limited tests. There is great necessity to be able to predict the number of cycles until the component fractures in long-duration operations based on these data.

In literature one can find an empirical equation

$$(\sigma_a)^m \times N_f = const \tag{10.47}$$

where $m$ is a constant that depends on the material. One should determine the value of $m$ from fatigue tests. As

$$\ln N_f = -m \ln \sigma_a + \ln const \tag{10.48}$$

the $m$ value can be calculated as the angle of inclination of the straight line to the x-axis in the graph $\ln N_f - \ln \sigma_a$.

Curves for the calculation of the exponent $m$ for some titanium-based alloys are presented in Figure 10.31. It is obvious that the curves correspond to the linear dependence. However, under relatively low stresses this law does not hold. It is true that $N_f$ is greater than $const/\sigma^m$ where $m$ and $const$ have been determined previously under middle or relatively high stress amplitudes.

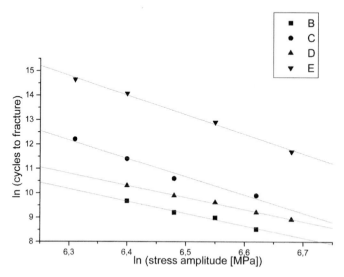

**Fig. 10.31** Dependence $\ln N_f - \ln\sigma_a$ for titanium-based alloys: B, non-treated material; C, as B after deep rolling; D, electropolished material; E as D after shot peening; B, C alloy Ti+6Al+7Nb, the fatigue test data of [61], D, E alloy Ti+6Al+4V, the test data of [63].

Table 10.5 presents the results for calculations of constants for titanium-based alloys.

**Table 10.5** Areas of applicability of (10.47) and constants in this equation for titanium-based alloys. Gas-turbine blades were tested for the alloy denoted as 5.

| No | Alloy | Processing | Measurement interval (MPa) | (10.4) is correct: | m | ln (const) | Test data |
|---|---|---|---|---|---|---|---|
| 1 | Ti+6Al+ +7Nb | No surface treatment | From 450 to 800 | From 600 to 800 | 4.95 | 42.02 | [77] |
| 2 | As 1 | Shot peening | From 550 to 800 | From 550 to 800 | 8.00 | 65.19 | |
| 3 | Ti+6Al+ +4V | No surface treatment | From 500 to 750 | From 500 to 750 | 5.07 | 42.13 | [64] |
| 4 | As 3 | Deep rolling | From 550 to 750 | From 550 to 750 | 7.50 | 59.94 | |
| 5 | Ti+7Al+ 2Mo+2Cr | Shot peening | From 500 to 760 | From 500 to 760 | 10.6 | 83.39 | [68] |

It is noteworthy that values of the power exponent are almost equal for different titanium-based alloys. The equation $(\sigma_a)^5 \times N_f = const$ is valid for the non-strengthened surface. In the case of shot peening or deep rolling the value of $m$ increases to 7.5–10.0. The number of cycles to fracture increases respectively.

Consequently, one may use (10.47) for a prediction of the fatigue life if the stress amplitude is not too small.

## 10.6
## Influence of Gas Adsorption

One should take into account the role of atmospheric conditions in the initiation and growth of surface fatigue cracks.

Figure 10.32 presents a model for fatigue-crack initiation on the surface. The model was originally proposed as early as 1956.

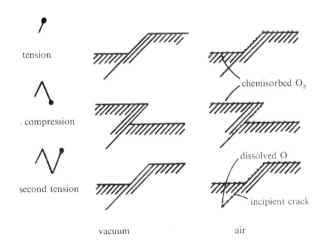

**Fig. 10.32** Fatigue-crack nucleation near a free surface by the effect of the single slip and due to environmental interactions. A model of Thomson, Wadsworth, Louat, and Neumann (after [71]).

The slip of dislocations along a slip plane during the phase of tensile loading produces steps at the surface. In reality, tens of thousands of one-signed dislocations generate the surface steps. The slip displacements and the surface steps can be diminished by reverse slip during unloading or subsequent compression loading. In an ideal vacuum or in an inert gas, the process would be completely reversible. In air conditions the step generation is also reversible, but only to some extent. In the real atmosphere the chemisorption of the embrittling atoms (such as O, N) or the formation of oxide layers make reverse slip impossible. During application of the alternating load the described process of the dislocation movement and persistent slip-band formation can gradually provide a roughening of the surface. In its turn this results in the appearance of stress concentrators and thereby facilitates crack initiation.

## 10.7 Summary

A considerable body of experimental evidence supports the general idea about the movement of deforming dislocations during the alternate loading of metals and alloys. The movement of mobile dislocations plays a significant role in fatigue failure preparation. A decrease in the electron work function during fatigue tests produces evidence that dislocations emerge at the surface and form surface steps. The accumulation of defects in the crystal lattice and the annihilation of defects during the beginning of cycling give way to each other. The process of fatigue damage is entirely reversible up to approximately 20–25 % of the lifetime. The reversibility of fatigue crack initiation during the first stage of fatigue is determined by thermodynamics.

The persistent slip bands are the cause of extrusions and intrusions at the surface. The slip bands can result in an initiation of cracks as well as in a temporary disappearance of these embryo cracks.

The density of dislocation dipoles in a metal increases as a result of cyclic loading. Discrete structures are developed in fatigued metals. Fatigue-crack initiation is interpreted as a state of dislocation structures, where the free energy is the sum of the elastic strain energy, the potential energy of the applied load and the surface energy of the crack. The elastic strain energy increases with the number of loading cycles. The accumulation of the elastic energy can be relieved by crack initiation along the persistent slip band.

The interaction of dislocations, the formation of pile-ups inside persistent slip bands, and the initiation of surface-crack embryos are underlying physical processes of fatigue mechanisms.

A crack must achieve a critical length $a_{incub}$ before it is able to grow. The crack embryo grows if

$$2\gamma S < F_1 + F_2$$

where $\gamma$ is the surface energy, $2S$ is the area of the crack embryo, $F_1$ is the free energy of dislocation pile-ups, and $F_2$ is the energy released as a result of the system relaxation and opening up a crack. The dependence of the free energy on the cycle number is a curve with a maximum. At this maximum $dF/dN = 0$. The greater the applied stress amplitude $\sigma_a$ the smaller the size of the crack embryo.

We obtain an expression for the length of the fatigue crack embryo of the form

$$a_{incub} = \frac{0.508 \gamma E}{(1 - \nu^2)(\sigma_a)^2} \qquad (10.49)$$

where $\gamma$ is the surface energy, $E$ is Young's modulus, $\nu$ is the Poisson coefficient, $\sigma_a$ is the stress amplitude. Consequently, the length of the crack embryo that can grow is inversely proportional to the stress amplitude squared.

It is reasonable to introduce a concept of the critical value of cycles $N_{incub}$ that corresponds to the critical crack length $a_{incub}$. If the number of cycles $N < N_{incub}$ the crack embryo does not grow. If the number of the loading cycles $N \geq N_{incub}$ the crack length begins to increase.

It is appropriate to divide the general number of cycles up to fracture $N_f$ into three periods:

$$N_f = N_{incub} + N_{gr} + N_{inst}$$

where $N_{incub}$ is the number of cycles when the crack length is less than the critical size, the crack does not grow, and the process is still reversible (an incubation period); $N_{gr}$ is the number of cycles when the fatigue crack grows in length (a period of growth); $N_{inst}$ corresponds to an instant fracture, when the section of a specimen or a component has decreased by the gradual growth of the crack (a period of fracture). Generally, $N_{inst} \ll (N_{incub} + N_{gr})$.

At large $\sigma_a$ $N_{incub} \ll N_{gr}$. The initiation of cracks is a very fast process under these conditions, and the crack begins to grow immediately.

On the other hand, under a relatively small applied stress amplitude the incubation period takes a considerable part of the specimen's life. The number of cycles until specimen fracture becomes very large and sometimes cannot be reached.

Thus, the fatigue-crack embryo or embryos originate at the surface, then the crack embryo begins to grow. The movement of vacancies to the tip of the crack plays a dominant role in crack growth.

The number of vacancies, which are generated by mobile intersecting screw components of dislocations, is much greater than the equilibrium number of vacancies that are formed spontaneously.

The external cycling stress at the crack tip creates a field of tensile stresses. The gradient in the tensile stress ensures a driving force for the relay-race diffusion of vacancies in the direction of the crack tip. Vacancies are attracted to the tension area in the tip of the crack. If a vacancy reaches a crack, it falls down inside it. The crack is a trap for the flow of vacancies. The replacement of an atom with a vacancy breaks the interatomic bonds in the tip of crack.

The rate of crack growth is given by

$$\frac{da}{dN} = \frac{b V_v}{\nu U_v} g S a \qquad (10.50)$$

where $a$ is the crack length, $N$ is the number of cycles, $b$ is the modulus of the Burgers vector, $V_v$ is the velocity of the vacancies, $U_v$ is the energy of vacancy generation, $\nu = dN/dt$ is the frequency of the fatigue tests, $g = \mathrm{grad}\,\sigma$, and $S$ is the area of the crack tip. Equation (10.50) is applicable if the size of the crack $a \geq a_{incub}$.

The dependence of the crack length on time is given by

$$a = A\,\exp(Gt) \tag{10.51}$$

where $A$ and $G$ are constants.

Increasing the exponent coefficient $G$ causes a faster growth of the fatigue crack. $G$ decreases as a result of the surface compressive residual stresses. On the contrary, this coefficient increases as a consequence of induced tensile residual stresses. These changes in $G$ indicate that macroscopic residual stresses affect the stress gradient near the tip of the fatigue crack.

The tip of the fatigue crack looks like a needle point. The value of the crack tip has been estimated to be from 2.4 to 40.9 nm if the ratio $a_f/a_{incub}$ is assumed to change by seven order of magnitude. This means that the tip of the crack is evaluated to be of the order of 10–170 interatomic distances. Approximately from 10 to 170 interatomic bonds are simultaneously broken in front of the growing crack.

The larger the stress amplitude the greater is the stress gradient near the crack tip and the larger the quantity of vacancy jumps to the crack during the cycle. At relatively small applied stresses the product $\Gamma g S$ increases slowly with $\sigma_a$. At relatively large amplitudes this product increases exponentially, which means an exponential rate of crack growth.

The empirical equation $(\sigma_a)^5 \times N_f = const$ is found to be correct for different titanium-based alloys with a non-strengthened surface. In the case of shot peening or deep rolling of the surface the exponent value increases to 7.5–10.0. The time until component fracture increases respectively.

One may use the equation $(\sigma_a)^m \times N_F = const$ to predict the fatigue life if the stress amplitude is not too small.

# 11
# Improvement in Fatigue Performance

In this chapter we consider some techniques for improvement of fatigue performance for industrial alloys. Topics that we are going to discuss are:
- intermediate thermal treatment;
- processing of alloys by electric impulses;
- combined strengthening of components, which involve restoring, the severe surface plastic strain and the armoring of components by a coating.

## 11.1
**Restoring Intermediate Heat Treatment**

We have studied the effect of an intermediate heat treatment on the structure and fatigue of the EP479 superalloy. The concept is as follows.

Imperfections accumulate in the crystal lattice of a superalloy under the influence of external alternating stresses. While the density of the lattice defects (dislocations and vacancies) is below a certain limit we can make an attempt to annihilate, or at least to reduce, the defect concentration. Until the surface crack embryo exceeds a critical size $a_{incub}$, the crack can still be closed.

To make this possible we discontinue the operation or the fatigue test of the material and use an intermediate heat treatment. Such heat treatment is assumed to increase the lifetime of the specimen during continued operation.

Specimens of the EP479 superalloy under investigation are shown in Figure 2.18 5. Specimens have been subjected to hardening at a temperature of 1313 K (1040°C) by quenching in oil. Afterwards they have been annealed at 923 K (650°C), polished with diamond paste and finally annealed at 853 K (580°C) in a vacuum of $1.33 \times 10^{-3}$ Pa ($1 \times 10^{-5}$ torr) for four hours. Fatigue tests were conducted under stress amplitudes 343 and 441 MPa at a resonance frequency of 381 Hz.

We have tested some hardened and annealed specimens until fracture. Others have been tested to $10^5$ or $5 \times 10^6$ cycles and the tests were interrupted. This pilot batch was thermally treated during the test interval according to the procedure (these specimens were hardened, annealed and polished again). Later, the fatigue tests were continued.

*Strained Metallic Surfaces.* Valim Levitin and Stephan Loskutov
Copyright © 2009 WILEY-VCH Verlag GmbH & Co. KGaA, Weinheim
ISBN: 978-3-527-32344-9

Experimental data are shown in Table 11.1. The intermediate heat treatment actually increases the total lifetime of the specimens. The average time until fracture for treated specimens is double, and four times as much for stress amplitudes of 343 and 441 MPa, respectively.

**Table 11.1** Effect of intermediate heat treatment on the fatigue lifetime of the EP479 superalloy specimens, cycles.

| Stress amplitude (MPa) | Without heat treatment | Intermediate heat treatment |
|---|---|---|
| 343 | $(5.5 \pm 2.6) \times 10^6$ | $(13.6 \pm 5.9) \times 10^6$ |
| 441 | $(0.9 \pm 0.5) \times 10^6$ | $(3.4 \pm 2.9) \times 10^6$ |

An appreciable scattering of data is typical for fatigue tests. The deviation in the Student's distribution reaches more than 50%.

Systematic nondestructive measurements of the electron work function make it possible to observe processes of plastic strain during fatigue tests. Figure 11.1 illustrates the corresponding distribution of the charge relief for superalloys under investigation. The method indicates that processes of defect accumulation during fatigue develop non-uniformly. The initiation of a dangerous crack localizes in a certain place of the specimen surface (see curves denoted as ■ in Figure 11.1a and as $E$, $F$, $G$ in Figure 11.1b).

It can be easily seen that the surface fatigue damage in superalloys is advanced in a trouble spot, and the work function decreases correspondingly. The appropriate intermediate heat treatment can cause structure relaxation at such a spot.

## 11.2
## Effect of the Current Pulse on Fatigue

Electric current pulses were shown to influence the plastic properties of metals [91–93]. However, the physical nature of the so called electroplastic effect has not yet been studied sufficiently. In particular it is not known if the current pulses have a specific effect on the crystal lattice.

The aim of our study is to investigate how the power current pulses influence fatigue strength, as well as the surface stresses, and microstructure parameters of a titanium alloy.

Specimens of the shape of a single shovel (Figure 2.18 2) are used.

Specimens are pre-strengthened by ball-bearings in the ultrasonic field. The frequency of vibrations of the chamber walls is equal to 17.2 kHz and the amplitude is 45 μm. The time $t$ of treatment is 300, 600, or 900 s. Ten specimens of each kind are treated.

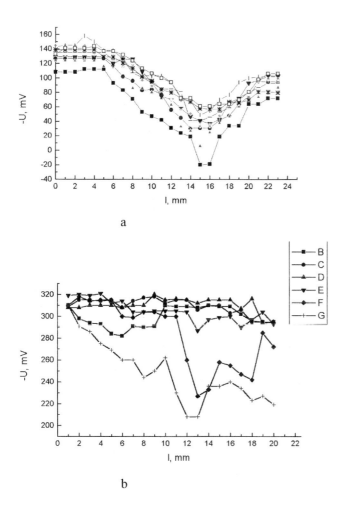

**Fig. 11.1** Increment in the contact potential difference along specimens of superalloys during fatigue tests: a, the EP479 superalloy after $N = 10^5$ cycles, measuring along 11 longitudinal lines of gauge length of the specimen; b, the EP866 superalloy after $N = 2 \times 10^5$ cycles, measuring along 6 lines, B – G.

An electric pulse is generated in the specimen from an electric capacitor battery of 200 µF. The current density is equal to 150 MA m$^{-2}$; the duration of the pulse is 50 µs. A light-beam oscillograph registers the pulses. Maximum temperatures that are measured by a thermocouple vary from 350 to 520 K.

After electric pulse treatment specimens are fatigue tested, followed by X-ray examination and for the measurement of the electric resistance. Fatigue tests are carried out at a frequency of 418 Hz. The stress amplitude was varied from 300 MPa to 600 MPa by 100 MPa after every $10^7$ cycles.

The results obtained are presented in Table 11.2.

**Table 11.2** The effect of electric pulse treatment (EPT) on fatigue life $N_f$, macroscopic residual stresses $\sigma$, microscopic stresses $\Delta a/a$, and the average subgrain size $D$ of the titanium-based VT3-1 alloy; $t$ is the time of specimen treatment by ball-bearings in the ultrasonic field.

| t (s) | $N_f$ ($10^7$ cycles) | | $\sigma$ (MPa) | | $\frac{\Delta a}{a}$ ($10^{-3}$) | | D (nm) | |
|---|---|---|---|---|---|---|---|---|
| | Before | EPT | Before | EPT | Before | EPT | Before | EPT |
| 300 | 2.5 ± 1.2 | 3.8 ± 0.5 | −466 | −311 | 1.5 | 2.6 | 12.4 | 10.5 |
| 600 | 3.5 ± 0.9 | 3.6 ± 0.6 | −594 | −437 | 1.2 | 2.4 | 12.3 | 9.2 |
| 900 | 2.9 ± 1.3 | 3.7 ± 0.7 | −648 | −447 | 1.5 | 2.0 | 15.0 | 11.8 |

It follows from the data that the pulse treatment improves fatigue strength. The improvement depends on the initial treatment of the metal surface. The largest increase in durability is observed after preliminary strengthening of specimens for 300 s. In this case, the number of cycles until fracture $N_f$ is about 50 % larger. The fatigue life after electric pulse treatment is in fact the same for all three specimen groups and equals $3.7 \times 10^7$ cycles (see the third column in Table 11.2).

The scattering of data is calculated according to the Student $t$-distribution. An important result is that the mean value of the durability scattering decreases after the electric pulse treatment from 40 to 16 %. The results of fatigue tests become more uniform under the influence of a current pulse.

It is of interest to compare the influence of electric pulse treatment with the heat effect of the direct current. A group of specimens was heated from 300 K to 520 K for 60 s. We obtained the following result: the mean fatigue life of the specimens was equal to $2.7 \times 10^7$ before and $2.6 \times 10^7$ cycles after this treatment. Hence the direct current has no positive effect on fatigue life.

Treatment by ball-bearings in the ultrasonic field creates compressive macroscopic residual stresses on the surface, so $\sigma$ is negative. One can see from the third column of Table 11.2 that the increase over time of the treatment with ball-bearings balls in the ultrasonic field creates surface stresses of −466 to −648 MPa. The pulse of the current decreases the absolute value of the macroscopic stress by 150 to 200 MPa. On the contrary, the microscopic stresses $\frac{\Delta a}{a}$ increase as a result of pulse treatment. The dimensions $D$ of the subgrains, that is, coherent diffracting areas decrease slightly.

The pulse treatment causes a decrease in the resistance of the alloy to direct electrical current. The drop is equal to 5–10 % and is greater than the error of measurement.

Fatigue is well known to be very sensitive to any surface defects, scratches, notches, dislocation pile-ups and cracks. From experimental data obtained we can conclude that the physical cause of the improvement in the fatigue properties is a local relaxation of the most stressed areas in the crystal lattice, particularly, at the surface.

Dislocation junctions, vacancies, cloud clusters and microscopic crack embryos in metals are in a non-equilibrium state because the regular atomic bonds are distorted or broken. Defects in the material are generally accepted to have an excess energy. If alternating forces are applied these active regions become sources of dislocation generation, the origination of pile-ups, an increase in the vacancy concentration and the initiation of microscopic cracks.

The current pulse passage is followed by heat release. Since any defect area has a greater electrical resistance than that of the regular crystal lattice, the heat of the electric current pulse is released at first in these stressed areas. Additional heat energy facilitates the transition of the structure to a state of further equilibrium. The decrease in the electrical resistance after the electric pulse treatment indicates unequivocally the presence of a relaxation process.

Macroscopic stresses, which are balanced throughout the bulk, decrease. However, the microscopic stresses $\Delta a / a$, which are balanced in microscopic areas (in crystallites), increase. The subgrains become smaller, so the structure of the specimen becomes more homogeneous.

An unusual property of electric pulses is the localization of heating in defect areas of the material. The electric pulse acts as a "directed repairer" of crystal lattice defects.

## 11.3
**The Combined Treatment of Blades**

We have investigated several groups of compressor blades. One can see the shape of a blade under study in Figure 8.5. The length of the blade was 80 mm. The material of which the blades were made was the titanium-based alloy VT 8M. The blades were selected after the non-failure operation time in a gas-turbine engine during 833, 952, and 1547 hours, respectively. The blades did not have any visible surface defects.

The blades are restored after selection by vibration polishing. This well-proven technology allows one to remove the smallest metallic particles and to form a favorable relief. Residual stresses at the surface are approximately $-350$ MPa and 40 µm throughout the depth as a result of vibration polishing.

After this procedure the blades are fatigued. The bases of fatigue tests were chosen to be equal to $20 \times 10^6$ or $100 \times 10^6$ cycles. We conduct the tests by means of a vibration table. The frequency of the natural oscillations of blades is 620–640 Hz. The fatigue test of a blade is conducted at a resonance mode up to the appearance a macroscopic crack of 1–3 mm length. By that time the frequency of oscillations decreases by 2–3%.

The fatigue test of blades is known to be an expensive procedure. One uses the standard method of tracing an $S - N$ curve for a final estimation of the effectiveness of the technology. However, the fatigue limit can be determined with reasonable accuracy using a so-called method of stair. One needs from 12 to 20 components in order to apply this method.

Figure 11.2 illustrates the data for fatigue tests of blades that were restored after the operation over 952 hours. The stress amplitude for a subsequent test depends upon the result of the previous test. One increases the stress amplitude if the blade stands the test for a given base. One decreases the stress amplitude if the blade fails. The stress step is ±20 or ±25 MPa.

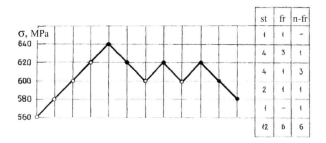

**Fig. 11.2** The sequence of fatigue tests of restored blades. The basis of the tests is equal to $20 \times 10^6$ cycles. $\sigma$ is the stress amplitude: ○, the blade has not failed; ●, the blade has failed; $st$ is the total sum of tested blades; $fr$ is the number of failed blades; $n - fr$ is the number of blades that have stood the test.

78% of failed blades have a fracture at the input edge. Far less, 22%, of failed blades have damage at the back. It is possible to express the fatigue limit, $\sigma_{fl}$, at the given base as

$$\sigma_{fl} = \sigma_0 + \left( \frac{\sum_{i=0}^{n} l_i}{\sum_{j=1}^{m} f_j} - \frac{1}{2} \right) \tag{11.1}$$

where $\sigma_0$ is the minimum stress amplitude (zero level of the stress), $l$ is the number of the test level, $n$ is the quantity of levels, $f$ is the number of blades that did not stand the test at all stress levels, $m$ is the quantity of failed blades. The mean value of $\sigma_{fl}$ is determined for a failure expectancy of 50% for the given test base.

The data obtained for fatigue parameters are presented in Table 11.3. The fatigue limit $\sigma_{fl}$ is of the order of 510–580 MPa. The variance in the fatigue limit in a group does not exceed 6%. Serial blades have a fatigue limit of 575 MPa on average and a variance of 9%. The coefficient of survivability $K_i$ has sufficiently high values (for the definition of $K_i$ see Section 8.2.2).

The variation in the fatigue limit is given by

$$v = \frac{S_\sigma}{\sigma_{fl}} \tag{11.2}$$

**Table 11.3** Data for the fatigue testing the compressor blades after operation, restoring and vibration polishing. The basis of the fatigue tests is $20 \times 10^6$ cycles.

| Fatigue parameters | Operation time (hours) | | |
| --- | --- | --- | --- |
| | 1547 | 518 | 583 |
| Fatigue limit $\sigma_{fl}$ (MPa) | 510 | 518 | 583 |
| Mean square of fatigue limit $S_\sigma$ (MPa) | 33 | 9 | 30 |
| Fatigue limit at 10% probability of fracture (MPa) | 470 | 507 | 553 |
| Coefficient of survivability $K_i$ | 0.655 | 0.405 | 0.860 |
| Variation $v$ | 0.064 | 0.017 | 0.051 |

The restored blades after vibration polishing reveal a fatigue resistance close to that for just-produced blades. Consequently, the process of restoration can be used for unimpaired blades.

After vibrational polishing the blades are subjected to two subsequent production operations. The first operation is the treatment of blades by ball-bearings in the ultrasonic field with the purpose of inducing residual compressive stresses. Further, one applies a finishing coating. One coats the surface of the blade feather by a thin layer of titanium nitride TiN. A scheme of the installation for covering of blades with a layer of titanium nitride is shown in Figure 11.3.

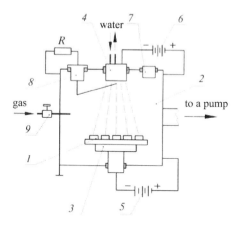

**Fig. 11.3** The installation for covering of blades with titanium nitride: 1, components; 2, the chamber; 3, the support; 4, the cathode; 5, the high-voltage rectifier; 6, the voltage source; 7, the anode; 8, the device for discharge initiation.

The process is to spray the TiN with an arc-heating plasma source. The process allows one to spray any conducting material. The ionization of plasma streams facilitates a fusion reaction for refractory compositions at a relatively low temperature (from 573 to 873 K). The high-speed plasmic stream ($10^3$ m s$^{-1}$) is condensed on the component surface.

The conditions for the covering process are as follows: the temperature equals to 823 K (550°C); the pressure is 240 Pa; the atmosphere in the chamber consists of 86% argon, 9% nitrogen, 6% hydrogen; the time for covering is 10 minutes.

We have investigated a different order for these work operations: vibrational polishing + coating; vibrational polishing + coating + treatment by vibrating ball-bearings; vibrational polishing + treatment by vibrating ball-bearings + coating. All blades under study were previously in operation in a gas-turbine engine for 800–1100 hours.

Figure 11.4 and Table 11.4 present the data obtained.

**Table 11.4** Data for testing of compressor blades after different treatments. The basis of the fatigue tests is $100 \times 10^6$ cycles.

| Processing | $\sigma_{fl}$ (MPa) | $S_\sigma$ (MPa) | $\sigma_{fl(p=10\%)}$ (MPa) | $K_i$ | $v$ |
|---|---|---|---|---|---|
| Vibration polishing + covering | 590 | 27 | 556 | 0.528 | 0.045 |
| Covering + treatment 5 min by ball-bearings | 630 | 17 | 608 | 0.622 | 0.027 |
| Covering + treatment 10 min by ball-bearings | 614 | 45 | 556 | 0.407 | 0.073 |
| Treatment 5 min by ball-bearings + covering | 654 | 27 | 620 | 0.618 | 0.041 |
| Treatment 10 min by ball-bearings + covering | 606 | 32 | 562 | 0.570 | 0.053 |

The strengthening of the blade surface by ball-bearings for 10 minutes results in lower properties ($\sigma_{fl} = 606$ MPa) than the strengthening for 5 minutes ($\sigma_{fl} = 654$ MPa). The relatively prolonged treatment of blades appears to cause an over-hardening of the thin edges of the blades.

The best fatigue performance of the blades ($\sigma_{fl} = 654 \pm 27$ MPa on the basis of $10^8$ cycles) is ensured by vibrational polishing, treatment by vibrating ball-bearings in the ultrasonic field for 5 minutes and then coating by titanium nitride.

## 11.4
### Structural Elements of Strengthening

A correlation is found between the fatigue behavior of blades and structure parameters.

The TiN coat has a greater specific volume than the titanium-based alloy. The covering of titanium nitride compresses the underlying layers of the blade. Therefore, the covering of the blade causes an increase in compressive residual stresses inside the surface layers of the alloy.

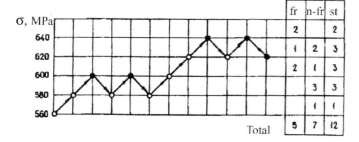

a

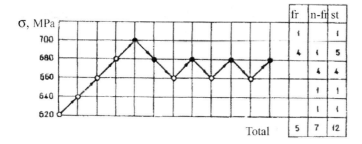

b

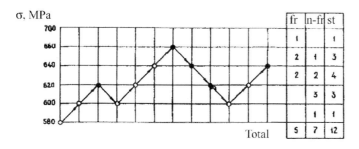

c

**Fig. 11.4** The sequence of fatigue tests for restored blades. The basis of tests is $100 \times 10^6$ cycles. $\sigma$ is the stress amplitude: ○, the blade has not failed; ●, the blade has failed; $st$ is the total sum of tested blades; $fr$ is the number of failed blades; $n - fr$ is the number of blades that have stood the test. $a$, operation for 833 hours, vibration polishing and covering by TiN; $b$, treatment by ball-bearings in the ultrasonic field for 300 s, and covering by TiN; $c$, treatment by ball-bearings in the ultrasonic field for 600 s, and covering by TiN.

In Figure 11.5 the points on the blade surface where X-ray diffractograms are taken are shown. Three points under study are situated on the back of the blade (1, 2, 3), and two points (4, 5) are chosen at the edges.

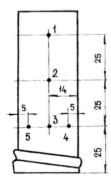

**Fig. 11.5** The scheme of the X-ray investigation of blades. 1, 2, 3, 4, and 5 are the points where diffractograms were taken; the other numbers are dimensions in mm.

The high-angle part of the X-ray diffractogram for a non-treated blade surface is presented in Figure 11.6. The diffractogram for a treated and coated blade is shown in Figure 11.7. The widening of the reflections indicates that residual stresses have increased sufficiently. Characteristic reflections of TiN (311) and (222) coincide with more intensive reflections of the alloy. Also, the layer of titanium nitride is too thin to observe diffraction patterns of TiN.

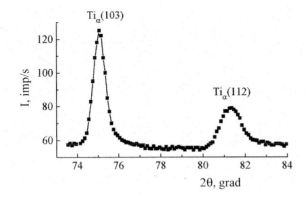

**Fig. 11.6** The diffractogram for a serial blade. Point 3 in Figure 11.5.

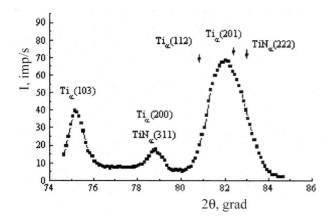

**Fig. 11.7** As in Figure 11.6 for a blade being treated by ball-bearings in the ultrasonic field and covered by titanium nitride.

**Table 11.5** Structural parameters of the blades after different types of surface treatment.

| Number | Processing | Point | $\sigma$ (MPa) | $\varepsilon\ (10^{-3})$ | $D$ (nm) |
|---|---|---|---|---|---|
| 1 | Operation in a gas-turbine engine for 1547 h | 1 | −360 | 5.5 ± 1.2 | 30 ± 9 |
|   |   | 2 | −155 |   |   |
|   |   | 3 | −105 |   |   |
|   |   | 4 | −110 |   |   |
|   |   | 5 | −140 |   |   |
| 2 | Vibration polishing + treatment by ball-bearings for 5 min in USF | 1 | −430 | 5.2 ± 1.5 | 85 ± 28 |
|   |   | 2 | −420 |   |   |
|   |   | 3 | −515 |   |   |
|   |   | 4 | −250 |   |   |
|   |   | 5 | −410 |   |   |
| 3 | Vibration polishing + treatment by ball-bearings for 10 min in USF | 1 | −390 | 2.8 ± 0.4 | 105 ± 37 |
|   |   | 2 | −360 |   |   |
|   |   | 3 | −570 |   |   |
|   |   | 4 | −420 |   |   |
|   |   | 5 | −470 |   |   |
| 4 | As 2 + covering by TiN | 1 | −430 | 3.5 ± 0.7 | 60 ± 34 |
|   |   | 2 | −710 |   |   |
|   |   | 3 | −780 |   |   |
|   |   | 4 | −600 |   |   |
|   |   | 5 | 570 |   |   |
| 5 | As 3 + covering by TiN | 1 | −695 | 5.1 ± 0.6 | 60 ± 29 |
|   |   | 2 | −760 |   |   |
|   |   | 3 | −720 |   |   |
|   |   | 4 | −605 |   |   |
|   |   | 5 | −635 |   |   |

Table 11.5 presents data for the structural investigations. One can draw the following conclusions from the data obtained.
- The level of macroscopic residual stresses at the blade surface decreases during the operating time.
- The value of compressive residual macroscopic stresses after treatment is larger on the back of the blade (points 2, 3) than it is on thick edges or at the feathered edge (points 4, 5, 1).
- The covering of a blade by titanium nitride results in an increase in residual macroscopic stresses by approximately $-200$ to $-300$ MPa.
- The microscopic residual stresses at the surface are of the order of $(2.8–5.5) \times 10^{-3}$. No relationship is discovered between technological processing and microscopical parameters.
- The dimensions of grains (subgrains) are found to be of the order of 30–100 nm. These values are determined with an error of up to 50%.

Figure 11.8, curve 2, shows an optimum alternative for the strengthening of blades by ball-bearings that vibrate in the ultrasonic field, and subsequent covering by the TiN layer. It provides compressive residual stresses of $-650$ MPa at the surface and their distribution to 120 μm in depth.

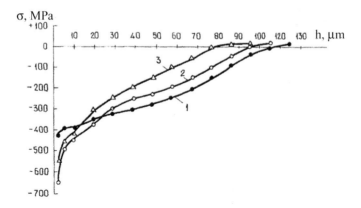

**Fig. 11.8** The distribution of residual stresses with depth for compressor blades: 1, treatment for 10 min by ball-bearings in the ultrasonic field; 2, as 1 and covering with titanium nitride; 3, covering with titanium nitride and then treatment for 10 min by ball-bearings in the ultrasonic field.

## 11.5
## Summary

The intermediate heat treatment favorably affects the fatigue life and structure of the EP479 superalloy.

While the density of crystal lattice defects during cycling is below a certain limit we can make an attempt to annihilate or at least to reduce the defect concentration. It is possible to discontinue the test or the operation of the material for this purpose and to use an intermediate heat treatment. Intermediate hardening and annealing is found actually to increase the total lifetime of specimens.

Electric pulse treatment increases the fatigue lifetime of a titanium-based alloy by 25–50 %. Macroscopic residual stresses decrease while microscopic stresses increase. The increase in the electrical resistance of specimens is evidence of a relaxation process. Heat release is found to occur at crystal lattice defects. Electric pulse advancement may be considered a promising method of alloy treatment.

We have investigated several groups of gas-turbine blades. The material of the blades is titanium-based VT 8M alloy. The blades are selected after non-failure operation time in a gas-turbine engine. They have no visible surface defects. The blades are restored after operation by vibration polishing. After this procedure the blades are fatigued. The bases of fatigue tests are chosen to be $20 \times 10^6$ or $100 \times 10^6$ cycles.

The fatigue limit can be determined with reasonable accuracy using a so-called stair-method. One needs from 12 to 20 components in order to use this method.

The restored blades, after vibration polishing, recover a fatigue strength close to that of just-produced blades. The process of restoration can be used for unimpaired blades. The fatigue limit $\sigma_{fl}$ is of the order of 510–580 MPa. The variance in the fatigue limit in a group does not exceed 6%. Serial blades have a fatigue limit of 575 MPa on average and a variance of 9%. The coefficient of survivability $K_i$ has sufficiently high values.

After vibrational polishing the blades are subjected to two subsequent production procedures. The first procedure is the treatment of blades by ball-bearings in the ultrasonic field with the purpose of inducing residual compressive stresses. Later one applies a finishing coating. One coats the surface of the blade feathers with a thin layer of titanium nitride TiN. The process involves spraying the TiN with an arc-heating plasma source. Ionization of the plasma streams facilitates to a fusion reaction for refractory compositions at a temperature of 573 to 873 K. The high-speed plasmic stream ($10^3$ m s$^{-1}$) is condensed on the component surface.

A different order for these three operations has been investigated: vibrational polishing + coating; vibrational polishing + coating + treatment by vibrating ball-bearings; vibrational polishing + treatment by vibrating ball-bearings + coating. The basis of the fatigue tests is $100 \times 10^6$ cycles.

The best fatigue performance of the blades ($\sigma_{fl} = 654 \pm 27$ MPa) is ensured by vibrational polishing, treatment by vibrating ball-bearings and then coating the feather with titanium nitride.

The value of compressive residual macroscopic stresses after treatments is larger on the back of the blade than it is on the thick edges or at the feathered edge. The coating of a blade causes an increase in compressive residual stresses inside the surface layers approximately by $-100$ to $-200$ MPa.

Microscopic residual stresses are found to be of the order of $(2.8 - 5.5) \times 10^{-3}$. The subgrain size is determined to be of the order of 30–100 nm.

# 12
# Supplement I

## 12.1
## List of Symbols

### 12.1.1
### Roman Symbols

| | | |
|---|---|---|
| $a$ | [m] | length of fatigue crack |
| $a_{incub}$ | [m] | length of equilibrium crack |
| $a_f$ | [m] | length of fraction crack |
| $A$ | [m²] | surface area |
| $A_a$ | [m²] | actual area of contact |
| $A_c$ | [m²] | contour area of contact |
| $\vec{b}$ | [m] | Burgers vector |
| $b$ | [deg] | width of X-ray reflection |
| $c_0$ | [–] | equilibrium concentration of vacancies |
| $c_{in}$ | [–] | concentration of vacancies due to dislocation intersections |
| $c_v$ | [–] | concentration of vacancies |
| $d$ | [m] | interplanar spacing |
| $da/dt$ | [m s$^{-1}$] | crack growth rate |
| $D$ | [m] | subgrain size |
| $dn$ | [–] | number of vacancies that fell into a crack |
| $e$ | [C] | electron charge |
| $E_F$ | [J] | Fermi energy |
| $E_j$ | [J] | energy of electron gas in jellium model |
| $E_M$ | [J] | Madelung energy |
| $E_{ps}$ | [J] | energy of electron – ion interactions |
| $E_s$ | [J] | adhesion energy |
| $F$ | [J] | free energy |
| $F$ | [N] | friction force |

*Strained Metallic Surfaces*. Valim Levitin and Stephan Loskutov
Copyright © 2009 WILEY-VCH Verlag GmbH & Co. KGaA, Weinheim
ISBN: 978-3-527-32344-9

| | | |
|---|---|---|
| $g$ | [Pa m$^{-1}$] | stress gradient at crack tip |
| $G_I$ | [J m$^{-2}$] | energy release rate |
| $h$ | [m] | depth of residual stress distribution |
| $h_{el}$ | [m] | elastic approach during indentation |
| $k$ | [J at$^{-1}$] | Boltzmann constant |
| $|\vec{k}|$ | [m$^{-1}$] | magnitude of wave vector |
| $k_F$ | [m$^{-1}$] | Fermi momentum |
| $K$ | [MPa m$^{\frac{1}{2}}$] | stress intensity factor |
| $K_i$ | [–] | survivability coefficient |
| $K_I$ | [MPa m$^{\frac{1}{2}}$] | stress intensity factor, mode I |
| $K_{th}$ | [MPa m$^{\frac{1}{2}}$] | threshold stress intensity factor |
| $l$ | [m] | length of dislocation pile-up |
| $l$ | [m] | mean length of dislocation segment |
| $L$ | [m] | crystal size |
| $n$ | [m$^{-1}$] | density of surface steps |
| $\bar{n}$ | [m$^{-3}$] | concentration of electrons in metal |
| $n_a$ | [m$^{-2}$] | density of adsorbed atoms |
| $n_{pu}$ | [–] | number of dislocations in pile-up |
| $n_+$ | [m$^{-3}$] | electron concentration |
| $n_1(z), n_2(z)$ | [–] | test functions of electron distributions |
| $n_a(\varepsilon)$ | [–] | number of electrons bonding surface atom with nearest neighbors |
| $n_b(\varepsilon)$ | [–] | number of electrons bonding surface atom with next-nearest neighbors |
| $N$ | [–] | number of cycles |
| $N_f$ | [–] | number of cycles until fracture |
| $N_{gr}$ | [–] | number of cycles during the crack grow |
| $N_{incub}$ | [–] | number of cycles until crack does not grow |
| $N_{inst}$ | [–] | number of cycles during instant fracture |
| $P$ | [N] | load |
| $P$ | [C · m] | dipole moment |
| $p_0$ | [C · m] | dipole moment of adsorbed atom |
| $r$ | [m] | radius of indenter |
| $r_1$ | [m] | radius of print curvature |
| $r_a$ | [m] | Pauling atomic radius |
| $r_s$ | [a.u.] | characteristic radius |
| $R$ | [Ω] | contact electric resistance |
| $R$ | [–] | load ratio |
| $R_1, R_2$ | [m] | distances between the nearest and next-nearest neighbors |
| $R_a$ | [m] | average height of peaks |

| | | |
|---|---|---|
| $R_z$ | [m] | arithmetic average deviation in profile |
| $S$ | [m$^2$] | cross-section of vacancy flow |
| $S_\sigma$ | [MPa] | mean square error of fatigue limit, scattering of durability |
| $t$ | [s] | time |
| $T$ | [K] | temperature |
| $U_0$ | [J at$^{-1}$] | energy activation of dislocation motion |
| $U_v$ | [J at$^{-1}$] | energy activation of vacancy generation |
| $v$ | [-] | variation in fatigue limit |
| $V$ | [m s$^{-1}$] | dislocation velocity |
| $V_v$ | [m s$^{-1}$] | vacancy velocity |
| $V_0$ | [m$^3$] | volume of unit cell |
| $V_0$ | [m s$^{-1}$] | pre-exponential factor in equation for $V$ |
| $V_n, V_{nn}$ | [-] | number of nearest and next-nearest neighbors of atoms |
| $W$ | [m] | model specimen size |

## 12.1.2
## Greek Symbols

| | | |
|---|---|---|
| $\alpha$ | [m$^3$] | polarizability of molecules |
| $\gamma$ | [J m$^{-2}$] | surface energy |
| $\gamma$ | [MPa] | applied shear stress |
| $\gamma$ | [m$^3$] | activation volume |
| $\Gamma$ | [-] | mean number of vacancy jumps per cycle |
| $\delta$ | [m$^{-1}$] | factor of dislocation multiplication |
| $\frac{\Delta a}{a}$ | [-] | microscopic residual stresses |
| $\Delta K$ | [MPa m$^{\frac{1}{2}}$] | range of stress intensity factor |
| $\varepsilon$ | [-] | strain |
| $\eta$ | [kg m$^2$ K$^{-1}$] | reduced amplitude of atomic vibrations |
| $\theta$ | [grad] | Bragg angle |
| $\lambda$ | [m] | wavelength |
| $\mu$ | [J at$^{-1}$] | chemical potential |
| $\mu$ | [–] | friction coefficient |
| $\mu_s, \mu$ | [MPa] | shear modulus |
| $\mu$ | [$\frac{C \cdot m}{m}$] | ratio dipole momentum to interatomic distance |
| $\nu$ | [–] | Poisson coefficient |
| $\nu$ | [s$^{-1}$] | frequency of cycling |
| $\nu_D$ | [s$^{-1}$] | Debye frequency |
| $\rho$ | [m$^{-2}$] | density of mobile dislocations |
| $\rho$ | [m] | radius of imprint |

| | | |
|---|---|---|
| $\sigma$ | [MPa] | residual macroscopic stresses |
| $\sigma_a$ | [MPa] | stress amplitude |
| $\sigma_d$ | [MPa] | fatigue limit in bending test |
| $\sigma_{fl}$ | [MPa] | fatigue limit |
| $\sigma_{min}$ | [MPa] | minimal value of residual stress |
| $\sigma_{sur}$ | [MPa] | residual stress at surface |
| $d\sigma/dx$ | [Pa m$^{-1}$] | stress gradient at crack tip |
| $\tau_1$ | [MPa] | shear applied stress |
| $\tau_1^D$ | [MPa] | stress created by dislocation pile-up |
| $\tau_d$ | [MPa] | fatigue limit in torsion test |
| $\tau_m$ | [MPa] | amplitude of shear stress |
| $\tau_s$ | [MPa] | threshold stress of dislocation slip |
| $\tau_{s0}, \tau_{f0}$ | [MPa] | start, final times of dislocation motion within cycle |
| $\tau_{th}$ | [MPa] | threshold stress of dislocation slip |
| $\varphi$ | [eV] | work function |
| $\psi$ | [deg] | angle between normals to surface and to reflecting plane |
| $\psi$ | [–] | wave function |
| $\omega$ | [rad s$^{-1}$] | angular frequency of cycling |

# 13
# Supplement II

## 13.1
### Growth of a Fatigue Crack. Description by a System of Differential Equations

The physical model of fatigue can be described by means of mathematical apparatus. The approach is analogous to that found in Chapter 7. We use a system of differential equations to trace the development of physical values during the influence of alternating stresses upon the material. In this connection we follow the physical mechanism of fatigue that has been developed and considered in Chapter 10.

Six structure parameters are of interest from this point of view. These are: $\tau_s, \rho, V, c_v, g$, and $a$ values.

### 13.1.1
### Parameters to be Studied

The change in the threshold stress for the dislocation motion is expressed as, (see (7.2))

$$\frac{d\tau_s}{dt} = \frac{\mu_s b}{4\pi} \cdot \left(\frac{n_{pu}}{\rho}\right)^{\frac{1}{2}} \cdot \frac{d\rho}{dt} \tag{13.1}$$

where $\mu_s$ is the shear modulus, $b$ is the Burgers vector, $n_{pu}$ is the number of dislocations in the pile-up and $\rho$ is the density of dislocations.

The dislocation density variation is given by, (see (7.3))

$$\frac{d\rho}{dt} = \delta \rho V \tag{13.2}$$

where $\delta$ is a coefficient of multiplication of mobile dislocations.

The dislocation velocity can be expressed as, (see (7.5))

$$\frac{dV}{dt} = \frac{V_0 \gamma}{kT} \exp\left[-\frac{U_0 - \gamma(\tau_m|\sin \omega t| - \tau_s)}{kT}\right] \left[\omega \tau_m |\cos \omega t| + \frac{\delta \mu_s b V (n_{pu}\rho)^{\frac{1}{2}}}{4\pi}\right] \tag{13.3}$$

*Strained Metallic Surfaces.* Valim Levitin and Stephan Loskutov
Copyright © 2009 WILEY-VCH Verlag GmbH & Co. KGaA, Weinheim
ISBN: 978-3-527-32344-9

where the pre-exponential factor, $V_0$, is estimated as [21]

$$V_0 = \frac{\nu_D b^2}{l} \tag{13.4}$$

where $\nu_D = 10^{13}$ s$^{-1}$ is the Debye frequency and $l = 1/\sqrt{\rho} \approx 10^{-6}$ m is the mean length of dislocation segments. The activation volume $\gamma = b^3$.

The increase in the relative vacancy concentration in the crystal lattice is proportional to the density of both intersecting dislocation systems, and the dislocation velocity multiplied by the volume $Q$. Thus we can write

$$\frac{dc_v}{dt} = \frac{1}{4}\rho^2 V Q \tag{13.5}$$

We find the equation for the stress gradient changing near the crack tip by differentiating (10.36)

$$\frac{dg}{dt} = -0.112 \times 10^{-3} g_0 \frac{da}{dt} \tag{13.6}$$

The rate of the crack growth is given by (10.21):

$$\frac{da}{dt} = \frac{b^2 v}{U_v} a g \Gamma S \tag{13.7}$$

Consequently, we have the system of six differential equations (13.1), (13.2), (13.3), (13.5), (13.6), (13.7) with six unknown quantities. The method of solution of the system was detailed in Chapter 7.

### 13.1.2
**Results**

We assume the following real and reasonable values for the physical parameters:

$b = 2.48 \times 10^{-10}$ m ($\alpha$ – Fe);     $\mu_s = 8.3 \times 10^{10}$ Pa;

$T = 293$ K;     $\tau_m = 68.9 \times 10^6$ Pa;

$\omega = 144.5$ rad s$^{-1}$;     $\delta = 2 \times 10^4$ m$^{-1}$;

$\Gamma S = 1.87 \times 10^{-18}$ m$^2$;     $M = 1.1 \times 10^{-4}$ m$^{-1}$;

$Q = 1.87 \times 10^{-28}$ m$^3$;     $U_v = 1.92 \times 10^{-19}$ J at$^{-1}$;

$V_0 = 0.818$ m s$^{-1}$;     $U_0 = 2.56 \times 10^{-20}$ J at$^{-1}$.

Initial values are assumed to be equal to $a = 0.5$ mm; $\rho = 1 \times 10^{12}$ m$^{-2}$; $V = 1.78 \times 10^{-8}$ m s$^{-1}$; $\tau_s = 37.6$ MPa; $c_v = 10^{-11}$; grad $\sigma = 4.5 \times 10^{12}$ Pa m$^{-1}$.

Data on the comparison of the test results and solutions of system of differential equations for fatigue-crack growth in low-carbon steel are shown in Figure 13.1. It can be easily seen that the calculated curves fit the experimental data very well.

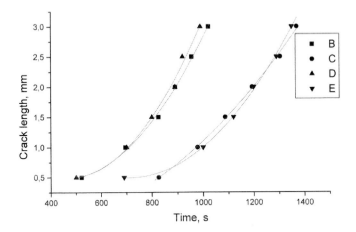

**Fig. 13.1** Crack length versus time of fatigue test for low-carbon steel. The stress amplitude is equal to 170 MPa, the frequency is 23 Hz. B, the non-cold-worked steel, fatigue test data [66]; D, data for solution of system differential equations for the same non-cold-worked steel; C, low-carbon steel after stretching of 340 MPa, the outcome of [66]; E, data for solution of system differential equations for pre-stretched steel.

The results for fatigued aluminum ($b = 2.86 \times 10^{-10}$ m, $\mu_s = 2.70 \times 10^{10}$ Pa) are presented in Figures 13.2–13.5.

Following initial values of the physical parameters have been assumed for the calculations.

$\rho = 1 \times 10^{12}$ and $2 \times 10^{12}$ m$^{-2}$;

$\mathrm{grad}\sigma = 1 \times 10^{12}$ and $2 \times 10^{12}$ Pa m$^{-1}$;

$V = 1.78 \times 10^{-8}$ m s$^{-1}$; $c_v = 10^{-11}$; $a = 10^{-4}$ m $= 0.1$ mm.

We can deduce the following conclusions from the data obtained.
- The model adequately represents physical processes during fatigue.
- During the cycling of metal the dislocation density increases by a factor of 3–6. The vacancy concentration increases by three orders of magnitude.
- The stress gradient near the crack tip considerably affects the rate of fatigue-crack growth. An increase of double in the stress gradient leads to a growth in the crack length of one order of magnitude.

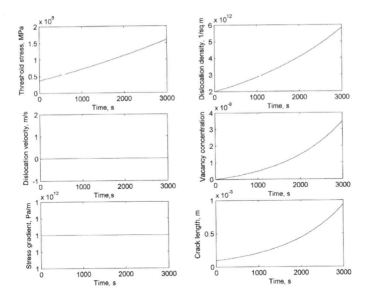

**Fig. 13.2** Variation of parameters for fatigued aluminum: description of processes by a system of six differential equations. The stress amplitude is 46 MPa, the frequency is 936 Hz, the number of cycles is $2.8 \times 10^6$. The stress gradient at the crack tip is equal to $1 \times 10^{12}$ Pa m$^{-1}$, the initial dislocation density is $2 \times 10^{12}$ m$^{-2}$. The crack grows from 0.1 to 1 mm, see the graph to the right below.

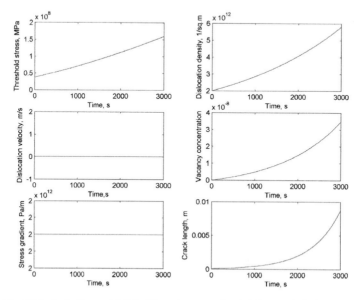

**Fig. 13.3** As in Figure 13.2, but the stress gradient at the crack tip is $2 \times 10^{12}$ Pa m$^{-1}$. The crack grows from 0.1 to 10 mm.

## 13.1 Growth of a Fatigue Crack. Description by a System of Differential Equations

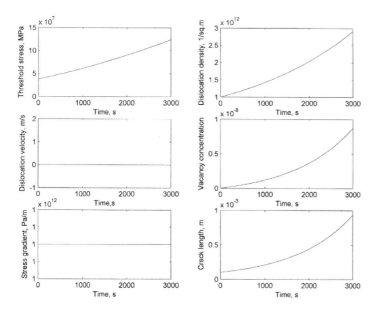

**Fig. 13.4** As in Figure 13.2, but the initial dislocation density is $1 \times 10^{12}$ m$^{-2}$, and the stress gradient at the crack tip is $1 \times 10^{12}$ Pa m$^{-1}$. The crack grows from 0.1 to 1 mm.

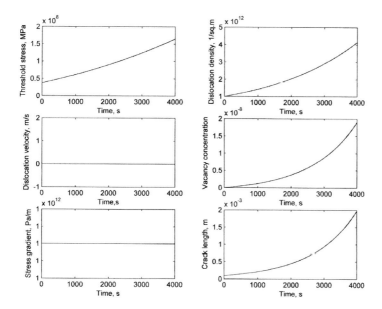

**Fig. 13.5** As in Figure 13.2, but the test time is 4000 s, that is, $3.7 \times 10^6$ cycles, the initial dislocation density is $2 \times 10^{12}$ m$^{-2}$. The crack grows from 0.1 to 2 mm.

# References

**1** J. M. Blakely, *Introduction to the Properties of Crystal Surfaces. International Series on Materials Science and Technology, Vol.12*, Pergamon Press, Oxford, **1973**.

**2** M. Prutton, *Surface Physics*, Second edition, Oxford Physics Series, Clarendon Press, Oxford, **1983**.

**3** A. Zangwill, *Physics at Surfaces*, Cambridge University Press, Cambridge-New York, **1988**.

**4** H. Lüth, *Surfaces and Interfaces of Solids*, Springer-Verlag, Berlin New-York, **2003**.

**5** A. Kiejna, K. F. Wojciechowski, *Metal Surface Electron Physics*, Elsevier Science, Oxford, **1996**.

**6** F. Bechstedt, *Principles of Surface Physics*, Springer-Verlag, Berlin-Heidelberg-New York, **2003**.

**7** J. Venables, *Introduction to Surface and Thin Film Processes*, Cambridge University Press, Cambridge, **2005**.

**8** D. Marx, *Surfaces and Contact Mechanics*, Center for Advanced Friction Studies, **2005**. Web version.

**9** R Shuttleworth, The surface tension of solids, *Proc. Phys. Soc.*, A63, (**1950**), p. 444.

**10** J. G. Che, C. T. Chan, W.-E. Jian, T. C. Leung, Surface atomic structures, surface energies, and equilibrium crystal shape of molybdenum, *Phys. Rev.*, B57, (**1998**), p. 1875

**11** L. Vitos, A. V. Ruban, H. L. Skriver, J. Kollar, The surface energy of metals, *Surf. Sci.*, 411, (**1998**), p. 186.

**12** P. Wynblatt, N. A. Gjostein, A calculation of relaxation, migration and formation energies for surface defects in copper, *Surf. Sci.* 12 (**1968**), p. 109.

**13** H. G. Kilian, V. I. Vettegren, V. N. Svetlov, Hierarchy of ensembles of defects on the strained surface in copper, *Phys. Sol.*, 43 (**2001**), p. 2107.

**14** B. Lang, R. W. Joyner, G. A. Somorjai, LEED-Studien des Hohen Index Crystal Oberflächen von Platin, *Surf. Sci.*, 30, (**1972**), p. 440.

**15** V. V. Levitin, *Atom Vibrations in Solids: Amplitudes and Frequencies*, Cambridge Scientific Publishers, **2004**.

**16** M. P. Shean, W. A. Dench, Compilation of experimental data determined with various electron energies for a large variety of materials, *Surf. Interf. Anal.*, 1, (**1979**), p. 1.

**17** R. W. James, *The Optical Principles of the Diffraction of X-Rays*, Bell and Sons LTD, London, **1950**.

**18** G. Maeder, X-ray diffraction and stress measurement, *Chemica Scripta*, 26A,(**1986**), p. 23.

**19** P. S.Prevèy, *X-ray Diffraction Residual Stress Techniques*, Metals Handbook, 10, 380, Metals Park: American Society for Metals (**1986**).

**20** B. E. Warren, X-ray studies of deformed metals, Progress in Metal Physics, 8, Pergamon Press, Ld-NY, (**1959**).

**21** J. Friedel, *Dislocations*, Pergamon Press, Oxford-London, **1964**.

**22** D. P. Woodruff, T. A. Delchar, *Modern Techniques of Surface Science*, Cambridge University Press, Cambridge-London-New York, **1986**.

**23** V. V. Levitin, S. V. Loskutov, M. I. Pravda, B. A. Serpetsky, A technique for measurement of strained state of specimens, Patent No 99084748, Ukraine, *J. Inventions*, (**2001**), p 10.

*Strained Metallic Surfaces*. Valim Levitin and Stephan Loskutov
Copyright © 2009 WILEY-VCH Verlag GmbH & Co. KGaA, Weinheim
ISBN: 978-3-527-32344-9

**24** W. Li, D. Y. Li, Effect of elastic and plastic deformations on the electron work function of metals during bending tests, *Phil. Mag.*, 84 (**2004**), p. 3717.

**25** W. Li, Y. Wang, D. Y. Li, Response of the electron work function to deformation and yielding behaviour of copper under different stress states, *Phys.stat.sol.*, 201 (**2004**), p. 2005.

**26** K. Besocke, B. Krahl-Urban, H. Wagner, Dipole moments associated with edge atoms; a comparative study of stepped Pt, Au and W surfaces, *Surf. Sci.*, 68 (**1977**), p. 39.

**27** M. Pivetta, F. Patthey, W.-D. Schneider, B. Delley, Surface distribution of Cu adatoms deduced from work function measurements, *Phys. Rev. B*, 65, 045417 (**2002**), p. 1.

**28** A. M. Kuznecov, Adsorption of water on metallic surfaces, *The Soros Edu. Journ.*, 6, (**2000**), p. 45.

**29** V. S. Fomenko, *Emission Properties of Elements and their Compounds*, Naukova Dumka, Kiev, **1980**.

**30** V. V. Levitin, S. V. Loskutov, V. V. Pogosov, The effect of strain and residual stresses in metals on the electron work function, *Phys. Met. Metall.*, 70, (**1990**), p. 73.

**31** N. W. Ashkroft, D. C. Langreth, Compressibility and binding energy of simple metals, *Phys. Rev.*, 155, 3, (**1967**), p. 682.

**32** P. M. Kobeleva, B. R. Gelchinsky, V. F. Uhov, On calculation of the surface energy of metals in the positive charge model, *Phys. Met. Metall*, 45, (**1978**), 1, p. 25.

**33** V. F. Uhov, R. M. Kobeleva, G. V. Dedkov, L. I. Termokov, *Electron Statistic Theory of Metals and Ionic Crystals*, Nauka, Moscow, **1982**.

**34** F. Jona, Re-examination of the structure of the clean surface of Al, *Solid State Phys.*, 10, (**1977**), 17, p. 619.

**35** F. Jona, D. Sondericher, P. M. Marcus, Al (111) revisited, *J. Phys. C*, 13, (**1980**), p. L155.

**36** A. Bianconi, R. C. Backrach, Al surface relaxation using surface extended X-ray absorbtion. Fine structure, *Phys. Rev. Lett.*, 79, (**1979**), p. 104.

**37** E. P. Giftopoulos, C. N. Ratsopulos, *The quantum thermodynamic definition of electronegativity and work function*, Proc. Second Int. Conf. on Thermion. Electrical Power Generation, Stresa, Italy, (**1968**), p.1249.

**38** V. I. Vettergren, V. L. Giliarov, S. Sh. Rahimov, V. N. Svetlov, Mechanism of the defect formation on the loaded surface in metals, *Phys.sol.*, 40 (**1998**), p. 668.

**39** L. A. Rudnitsky, Electron work function for nonideal surface of metals, II. Stepped surface and surface of microscopic granules, *J. Techn. Phys.*, 50, (**1980**), p. 355.

**40** L. A. Rudnitsky, Some surface and bulk properties of solids in terms of electronegativity, *J. Phys. Chem.*, 53, (**1979**), p. 3003.

**41** G. A. Malygin, Mechanism of formation of the nanometric deformation steps on the surface of plastic deformed metals, *Phys.sol.*, 43 (**2001**), p. 248.

**42** H. G. Kilian, V. I. Vettergren, V. N. Svetlov, Ensembles of defects on the strained surface of metals as a result of a reversible aggregation, *Phys. Sol.*, 42, (**2000**), p. 2024.

**43** R. I. Minc, V. P. Melehin, V. S. Kortov, U. D. Semko, The variation in the work function during tension of copper and aluminum, *Proc. Inst. High. Educ., Non-ferrous Met.*, (**1969**), p. 113.

**44** L. D. Landau, E. M. Lifshiz, *Theory of Elasticity*, Pergamon, Oxford, **1986**.

**45** J. H. Underwood, G. R. O'Hara, J. J. Zalinka, Analiysis of elastic-plastic, ball indentation to measure strength of high-strength steels, *Exper. Mechanics*, 26, (**1986**), p. 379.

**46** M. Scherge, D. Shakhvorostov, K. Pöhlmann, Fundamental wear mechanism of metals, *Wear* 255, (**2003**), p. 395.

**47** T. Kasai, X. Y. Fu, D. A. Rigney, A. L. Zharin, Applications of a non-contacting Kelvin probe during sliding, *Wear*, 225-229, (**1999**), p. 1186.

**48** I. Garbar, Microstructural changes in surface layers of metal during running-up friction processes, *Meccanica*, 36, (**2001**), p. 631.

**49** Z. B. Wang, N. R. Tao, S. Li, W. Wang, G. Liu, J. Lu, K. Lu, Effect of surface nanocrystallisazion on friction and wear properties in low carbon steel, *Mater. Sci. Eng.*, A352, (**2003**), p. 144.

50 Y. LI, D. Y. LI, Prediction of elastic-contact friction of transition metals under light loads based on their electron work function, *J. Phys. D:Appl. Phys.*, 40, (**2007**), p. 5980.

51 W. G. JOHNSTON, J. J. GILMAN, Dislocation velocities, dislocation densities, and plastic flow in lithium fluoride crystals, *J. Appl. Phys.*, 30 (**1959**), p. 129.

52 H. WIEDERSICH, A quantitative theory for the dislocation multiplication during the early stages of the formation of glide bands, *J. Appl. Phys.*, 33 (**1962**), p. 854.

53 L. MATOHNYUK, *Accelerated Fatigue Tests*, Naukova Dumka, Kiev **1988**.

54 S. A. TEUKOLSKY, W. T. VETTERLING, W. H. PRESS, B. P. FLANNARY, *Numerical Recipes in Fortran*, Cambridge University Press, Cambridge, **1999**.

55 V. LEVITIN, *High Temperature Strain of Metals and Alloys*, Wiley-VCH Verlag GmbH, Weinheim, **2006**.

56 A. V. KULEMIN, V. V. KONONOV, I. A. STEBELKOV, On a choice of an optimal regime of the material hardening in an ultrasonic field, *Phys. Chem. Mat. Treat.*, (**1982**), p. 93.

57 V. K. YATSENKO, G. Z. ZAYZEV, V. F. PRITCHENKO, L. I. IVSHENKO, *Increase in the Load-Carrying Ability of Components by the Diamond Smoothing*, Machine-Building, Moscow, **1985**.

58 G. V. PUHALSKA, A. D. KOVAL, L. P. STEPANOVA, Technological peculiarities of formation of the surface layer properties for compressor blades, *New Mater. Techn. Met. Mach.*, 1, (**2006**), p. 47.

59 M. A. S. TORRES, H. J. C. VOORWALD, An evaluation of shot peening, residual stress and stress relaxation on the fatigue life of AISI 4340 steel, *Int. J. Fatig.*, 24, (**2002**), p. 877.

60 G. A. WEBSTER, A. N. EZELIO, Residual stress distributions and their influence on fatigue lifetimes, *Int. J. Fatig.*, 23, (**2001**), p. S375.

61 L. WAGNER, Mechanical surface treatments on titanium, aluminum and magnesium alloys, *Mater. Sci. Eng.*, A263, (**1999**), p. 210.

62 T. ROLAND, D. RETRAINT, K. LU, J. LU, Fatigue life improvement through surface nanostructuring of stainless steel by means of surface mechanical attrition treatment, *Scripta Mater.* 54, (**2006**), p. 1949.

63 R. K. NALLA, I. ALTENBERGER, U. NOSTER, G. Y. LIU, B. SCHOLTES, R. O. RITCHIE, On the influence of mechanical surface treatments – deep rolling and laser shock peening – on the fatigue behaviour of Ti-6Al-4V at ambient and elevated temperatures, *Mater. Sci. Eng.*, A355, (**2003**), p. 216.

64 A. VINOGRADOV, S. HASHIMOTO, V. I. KOPYLOV, Enhanced strength and fatigue life of ultra-fine grain Fe-36Ni Invar alloy, *Mater. Sci. Eng.*, A355, (**2003**), p. 277.

65 H. GUECHICHI, L. CASTEX, Fatigue limits prediction of surface treated materials, *J. Mater. Proc. Techn.*, 172, (**2006**), p. 381.

66 M. N. JAMES, D. J. HUGHES, Z. CHEN, H. LOMBARD, D. G. HATTINGH, D. ASQUITH, J. R. YATES, P. J. WEBSTER, Residual stresses and fatigue performance, *Eng. Fail. Anal.*, (**2006**), p. 1.

67 J. D. ALMER, J. B. COHEN, B. MORAN, The effects of residual macrostresses on fatigue crack initiation, *Mat. Sci. Eng.*, A284, (**2000**), p. 268.

68 T. ROLAND, D. RETRAINT, K. LU, J. LU, Enhanced mechanical behavior of a nanocrystallised stainless steel and its thermal stability, *Mater. Sci. Eng.*, A 445-446, (**2007**), p. 281.

69 I. NIKITIN, I. ALTENBERGER, H. J. MAIER, B. SCHOLTES, Mechanical and thermal stability of mechanically induced near-surface nanostructures, *Mater. Sci. Eng.*, A403, (**2005**), p. 318.

70 U. KRUPP, *Fatigue Crack Propagation in Metals and Alloys*, Willey-VCH Verlag GmbH, Weinheim, **2007**.

71 S. SURESH, *Fatigue of Materials*, Cambridge University Press, Cambridge, **1998**.

72 G. R. IRWIN, Onset of fast crack propagation in high strength steel and aluminum *Proc. Sec. Sagamore Conf.*, 2, N-Y, Syracuse University, **1956**, p. 289.

73 P. C. PARIS, M. P. GOMEZ, W. P. ANDERSON, A rational analytic theory of fatigue, *The Trend. Eng.* 13, (**1961**), p. 9.

74 ASTM STANDART E-647, *Annual book of ASTM standards* 0.3.01, (**1990**), p. 642.

75 L. Bartosiewicz, A. R. Krause, A. Sengupta, S. K. Putatunda, Application of a new model for fatigue threshold in a structural steel weldment, *Eng. Frac. Mech.*, 45, (**1993**), p. 463.

76 M. Karimi, T. Roarty, T. Kaplan, Molecular dynamics simulations of crack propagation in Ni with defects, *Modelling Simul. Mater. Sci. Eng.*, 14, (**2006**), p. 1409.

77 D. L. Davidson, The growth of fatigue cracks through particulate SiC reinforced aluminum alloys, *Eng. Frac. Mech.*, 33, (**1989**), p. 965.

78 J. Schijve, Fatigue of structures and materials in the 20th century and the state of the art, *Int. J. Fatig.*, 25, (**2003**), p. 679.

79 J. P. Hirth, J. Lothe, *Theory of Dislocations*, Second Edition, Krieger, Malabar, Florida, **1992**.

80 Xu-Dong Li, L. Edwards, Theoretical modelling of fatigue threshold for aluminum alloys, *Eng. Frac. Mech.*, 54, (**1996**), p. 35.

81 T. Mura, A theory of fatigue crack initiation, *Mater. Sci. Eng.*, A176, (**1994**), p. 61.

82 D. M. Nissley, Thermomechanical fatigue life prediction in gas turbine superalloys: a fracture mechanics approach, *Amer. Inst. Aeron. Astron. Journ.*, 33, (**1995**), p. 1114.

83 W. W. Milligan, S. D. Antolovich, The mechanisms and temperature dependence of superlattice stacking fault formation in the single crystal superalloy PWA 1480, *Met. Trans. A*, 22A, (**1991**), p. 2309.

84 J. S. Koehler, On the dislocation theory of plastic deformation, *Phys. Rev.*, 60, (**1941**), p. 397.

85 A. H. Cottrell, D. Hull, Extrusion and intrusion by cyclic slip in copper, *Proc. Royal Soc.*, 242, (**1957**), p. 211.

86 M. Toparli, A. Özel, T. Aksoy, Effect of the residual stresses on the fatigue crack growth behavior at fastener holes, *Mater. Sci. Eng.*, A225, (**1997**), p. 196.

87 J. E. LaRue, S. R. Daniewicz, Predicting the effect of residual stress on fatigue crack growth, *Int. J. Fatig.*, 29, (**2007**), p. 508.

88 B. S. Bokstein, *Diffusion in Metals*, Metallurgy, Moscow, **1978**.

89 J. Schijve, Fatigue predictions and scatter, *Fatigue Fract. Eng. Mater. Struc.*, 17, (**1994**), p. 381.

90 S. V. Loskutov, *The effect of strain on the structure and on the energetic state of surface layers in metals*, PhD thesis, Institute for Metal Physics, NASU, Kiev, **2005**.

91 K. Okazaki, M. Kagawa, H. Conrad, A study of the electroplastic effect, *Scripta Metal.*, 16, (**1978**), p. 1063.

92 V. I. Spicyn, O. A. Troicky, *Electroplastic Deformation of Metals*, Nauka, Moscow, **1985**.

93 S. L. Livesay, X. Duan, R. Priestner, J. Colins, An electroplastic effect in 3.25 % silicon steel, *Scripta Metal.*, 44, (**2001**), 5, p. 803.

94 S. Mrowec, *Defects and Diffusion in Solids*, Elsevier Scientific Publishing Company, Oxford-N.-Y., **1980**.

95 V. V. Pogosov, V. V. Levitin, S. V. Loskutov, On strain-emission phenomenon in metals, *Let. J. Tech. Phys*, 16, (**1990**), p. 14.

96 François Cardarelli, *Materials Handbook*, Springer, London, **2007**.

97 W. Missol, Calculation of the surface energy of solid metals from work function values and electron configuration data, *Phys.stat.sol.(b)*, 58, (**1973**), p. 767.

# Index

## a
alloy
- aluminum-based   148–151, 159, 206
- EI698   56, 128
- EP479   56, 64, 161, 162, 219, 220
- EP866   56, 64, 188
- Invar   153, 155, 156
- nickel-based   55, 112
- PWA 1480   177, 178, 184
- titanium-based   55, 63, 104, 111, 112, 128ff., 137, 151, 160, 164
- – VT 3-1   56, 63, 92, 222
- – VT 8   56, 130, 134
- – VT 8M   56, 223ff.
- – VT 9   56, 128
- ZhS6K   56, 125

atomic units   20, 22, 75

## b
Bragg angle   45, 46, 127
Bragg formula   41

## c
chemical potential   7, 8
coefficient of survivability   133, 134, 224, 225, 231
compressor blades   128ff., 223ff.
- coating   225–226
- dimensions   129
- distribution contact potential difference   135, 136
- fatigue limit   133–136, 223–225
- macroscopic residual stresses   130–133, 144ff.
- – redistribution during operation   155, 157
- microscopic parameters   132, 133
- stair method of fatigue test   223–226
- subgrain size   132

computer simulation
- fatigue processes   115ff., 237ff.

conjugated pairs   89ff.
- electric conductivity   89–92

contact electrical resistance
- behavior during loading and unloading   88–90
- installation   50

contact interaction metallic surfaces   87ff.
contact potential difference   49, 92, 93, 135, 136
cycle number until fracture   176, 210

## d
deep rolling   151, 152
diffraction technique   33ff.
- electron scattering   33–35
- interference function   37
- LEED method   33–40
- measurement of macroscopic residual stresses   40–47
- – experimental installation   44ff.
- measurement of microscopic stresses   47
- patterns   34, 37, 162, 163
- reciprocal lattice   37–40
- RHEED method   40ff.
- X-ray studies   40ff., 125, 138, 228

disks   126–128
dislocation
- density   47, 109, 116, 119, 237–239
- – in fatigued aluminum   111
- – in fatigued Ti-based alloy   111
- intersections   195
- pile-ups   189–191, 216
- structure of fatigued superalloys   183
- velocity   237

distribution of chemical elements   137ff.

## e
electric current pulse   220–223
electron
- density   27–29
- distribution near surface   18, 24–30
- scattering   33–35
- subsystem parameters   22
- surface step distribution   26, 29

energy release rate   168, 171

## f

fatigue   103ff., 181ff., 219ff.
- computer simulation   115–121, 237–239
-- parameter evolutions   118ff.
-- system of differential equations   115ff.
- dislocation interaction   181, 184
- empirical and semi-empirical models   165ff.
- failure evolution   205–207
- life   135, 140, 147ff., 215, 220, 222
-- effect of surface treatment   147–153
-- prediction   171ff., 173ff.
- periods   192–194
- physical mechanism   181ff.
- prediction of location   103ff.
- reversibility   109, 112, 187, 188
- strength prediction   173–178
fatigue crack
- classification   168, 169
- dependence on physical parameters   195–198
- embryo size   188, 192, 207
- growth   195ff., 210, 212, 237–239
-- acceleration   202, 203
-- condition   192
-- rate   169, 170, 195–205, 237–239
-- vacancy mechanism   195–198
- in linear elastic fracture mechanics   166ff.
- in model crystal   171–172
- incubation length   206
- incubation period   192–194
-- dependence on stress amplitude   192, 193
- influence of gas adsorption   215
- influence of structure factors   143
- length   192ff., 201–203, 206
-- effect of cycling time   201
- needle-like tip   204, 210
- origination   148, 150, 181–187, 192
- prediction of location   104, 134–136
- propagation   169, 171, 192, 194
- stress gradient at tip   197, 204, 208, 209, 212, 237–239
- tip size   204, 210
fatigue limit   124, 126, 133, 134, 137, 141, 223–226
fatigue performance improvement   219ff.
- by combined treatment   219–226
- by intermediate heat treatment   219, 220
- by means of electric pulses   220–223
free electron model   18, 20–23
free energy   8–10, 188, 191, 192
friction
- coefficient   96–100
- effect on the work function   95–100
- surface structure   95–99

## g

gas adsorption   215
gas-turbine
- components   123–127, 128ff., 129
- surface residual stresses   124, 130, 131
- surface treatment   124, 127, 131, 133

## i

incubation period of fatigue crack   192
indentation of surface layers   53, 87ff.
intensity stress factor   167ff.
interference function   35, 37
intermediate heat treatment   219, 220
internal energy   7, 188
intrusions and extrusions   184–187

## j

jellium model   27, 73ff., 74
J-integral   176

## k

Kelvin technique   49–51

## m

mechanical indentation   87ff.
- elastic unloading   89
- elastic-plastic loading   87, 89
- Hertz problem   87
metals
- Al   59, 63, 71, 111, 117, 118
- Cu   59, 71, 184, 185
- Ni   27, 59, 71
- Ti   27
microstructure   95, 161
- stability   161–165
model of free electrons   20–23
modeling
- fatigue   115ff., 237ff.
-- changes in parameters   237
-- equations   115–118, 237–238
- work function   73ff., 76–82

## n

nanometric defects   16, 17, 81–82
nanostructuring of metal surface   143, 162

## p

Paris equation   168–171
persistent slip bands   186–187
physical mechanism
- fatigue   181ff.
-- crack growth   195, 237–239
-- crack origination   181, 184–192
- work function decrease   63, 65–66

## r

reduced amplitudes of atomic vibrations 23
relaxation 154ff.
residual stresses
– classification 41
– distribution in depth 47, 144–148, 158
– macroscopic 41–44, 131, 222, 230, 231
– microscopic 41, 133, 222, 230, 231
– redistribution under cycling 154ff., 156, 158, 160
– X-ray measurement 40ff.
– – installation 44–47

## s

S - N curves 150, 151, 154, 213–215
– for titanium-based alloys 214
Schrödinger equation 20
shot peening 131, 132, 134, 212
specimens for fatigue tests 56
steel
– 316L 150, 161, 163
– AISI 1080 159, 204
– AISI 304 164
– AISI 4340 148, 152, 154, 159
– low-alloy 174
– structural 161
strain-emission phenomenon 66
strained metallic surfaces 59ff., 143, 184ff.
– compressor blades 128ff., 130–133
– fatigue strength 150–153
– gas-turbine components 123ff.
– grooves of disks 126–128
– structure stability 161ff.
– turbine blades 124
stress intensity factor 167ff.
structure parameters
– affecting fatigue crack 143
– macroscopic 41, 222, 230
– microscopic 47, 222, 230
– physical models 115ff., 237ff.
structure relaxation 161ff., 220
subgrain size 132
superalloy
– EP479 64, 219–220
– EP866 64, 119, 120, 188
– MAR-M509 177
– PWA 1480 177, 178, 184
– ZhS6K 125
superstructure notation 13
surface
– crystal structure 11ff.
– defects 14ff.
– distribution of electrons 18
– – jellium model 27, 73–76
– distribution of residual stresses 144–148, 156
– energy 8–11, 208
– indentation 53, 87
– intrusions and extrusions 184–187
– nanometric defects 16, 17, 81, 82
– nanostructuring 162
– peculiarities 7ff.
– profile 144
– relaxation 11
– stress 9–11
– superstructure 13
surface steps 15–17, 65ff.
– density 65
– effect on work function 65ff.
– formation 15
– in Au and Pt 65
– in fatigued metals 118, 119
surface treatment 147–153
– by bearing balls in ultrasonic field 90, 125, 131, 134, 137, 145–148, 222
– by deep rolling 151, 152, 164, 214, 215
– by ion-plasma nitriding 146, 157
– by shot peening 131, 132, 134, 137, 145, 151–154, 212, 214, 215
– by vibratory polishing 134, 137, 154

## t

thermomechanical crack growth 176, 177
threshold stress 109
titanium nitride covering 225–226
turbine blades 124–126

## u

ultra-violet irradiation 51
– influence on work function 67, 68

## v

vacancy 14
– concentration 237–239
– displacement to crack tip 198
– generation 195
– number of jumps per cycle 210–213
– velocity 197, 200

## w

wave function 20, 24, 25
wear 95
work function 79
– above future fatigue crack 104–106
– definition 48–50
– dependence on
– – adsorption 67–70
– – cycling 105, 108, 187, 221
– – elastic strain 59–60
– – friction and wear 95–100

– – heating   67, 69
– – nanometric defects   82
– – plastic strain   61–64
– – roughness   90, 93, 101
– experimental installation   50, 51
– fatigue predict possibility   104, 134–136
– fatigued metals   104, 188, 220
– measurement   52
– modeling   73ff.
– – development of model   76, 78
– – elastic strained single crystal   73–76
– – influence of nanometric defects   81–82
– – neutral orbital electronegativity   78–81
– reversibility   108, 109, 188
– strained surfaces   61–64
– – for metals   61–64
– – for superalloys   64, 188
– – for titanum-based alloys   64
– – physical mechanism   65–66
– variation during fatigue tests   105, 108, 188, 220

**x**

X-ray measurement of residual stresses   40–44
– installation   44–47
X-ray reflection   46, 130, 132, 141, 165, 228
– shift   41
– width   165, 174